Set, Measure, and Probability Theory

RIVER PUBLISHERS SERIES IN MATHEMATICAL, STATISTICAL AND COMPUTATIONAL MODELLING FOR ENGINEERING

Series Editors:

Mangey Ram
Graphic Era University, India

Tadashi Dohi
Hiroshima University, Japan

Aliakbar Montazer Haghighi
Prairie View Texas A&M University, USA

Applied mathematical techniques along with statistical and computational data analysis has become vital skills across the physical sciences. The purpose of this book series is to present novel applications of numerical and computational modelling and data analysis across the applied sciences. We encourage applied mathematicians, statisticians, data scientists and computing engineers working in a comprehensive range of research fields to showcase different techniques and skills, such as differential equations, finite element method, algorithms, discrete mathematics, numerical simulation, machine learning, probability and statistics, fuzzy theory, etc.

Books published in the series include professional research monographs, edited volumes, conference proceedings, handbooks and textbooks, which provide new insights for researchers, specialists in industry, and graduate students.

Topics included in this series are as follows:-

- Discrete mathematics and computation
- Fault diagnosis and fault tolerance
- Finite element method (FEM) modeling/simulation
- Fuzzy and possibility theory
- Fuzzy logic and neuro-fuzzy systems for relevant engineering applications
- Game Theory
- Mathematical concepts and applications
- Modelling in engineering applications
- Numerical simulations
- Optimization and algorithms
- Queueing systems
- Resilience
- Stochastic modelling and statistical inference
- Stochastic Processes
- Structural Mechanics
- Theoretical and applied mechanics

For a list of other books in this series, visit www.riverpublishers.com

Set, Measure, and Probability Theory

Marcelo S. Alencar
Institute for Advanced Studies in Communications (Iecom)
Federal University of Rio Grande do Norte (UFRN), Brazil

Raphael T. Alencar
Raydiall Automotive, France

NEW YORK AND LONDON

Published 2024 by River Publishers
River Publishers
Alsbjergvej 10, 9260 Gistrup, Denmark
www.riverpublishers.com

Distributed exclusively by Routledge
605 Third Avenue, New York, NY 10017, USA
4 Park Square, Milton Park, Abingdon, Oxon OX14 4RN

Set, Measure, and Probability Theory / by Marcelo S. Alencar, Raphael T. Alencar.

© 2024 River Publishers. All rights reserved. No part of this publication may be reproduced, stored in a retrieval systems, or transmitted in any form or by any means, mechanical, photocopying, recording or otherwise, without prior written permission of the publishers.

Routledge is an imprint of the Taylor & Francis Group, an informa business

ISBN 978-87-7022-847-3 (hardback)
ISBN 978-87-7004-048-8 (paperback)
ISBN 978-10-0380-895-4 (online)
ISBN 978-10-3262-645-1 (master ebook)

While every effort is made to provide dependable information, the publisher, authors, and editors cannot be held responsible for any errors or omissions.

This book is dedicated to our family.

Contents

Preface	xiii
Acknowledgements	xv
List of Figures	xvii
List of Tables	xxiii
List of Abbreviations	xxv

1 Advanced Set Theory — 1
  1.1   Set Theory 1
  1.2   Basic Set Theory 2
  1.3   The Axioms of Set Theory 4
  1.4   Operations on Sets 5
  1.5   Families of Sets 7
      1.5.1   Indexing of Sets 10
  1.6   An Algebra of Sets 10
  1.7   The Borel Algebra 12
  1.8   Cardinality 13
      1.8.1   Equivalence of Sets 13
      1.8.2   Countable Sets 14
      1.8.3   Uncountable Sets 16
      1.8.4   Cardinality Properties 19
  1.9   Georg Cantor 20
  1.10 Problems 22

2 Relations and Functions — 23
  2.1   Definition of a Relation 23
      2.1.1   Relation Representation 24
      2.1.2   Types of Relations 26

2.2		Definition of Function	27
	2.2.1	Types of Functions	28
2.3		Mathematical Functions	31
	2.3.1	Indicator Function	33
	2.3.2	Fuzzy Sets	34
	2.3.3	Properties of Set Functions	36
2.4		The Count of Arts and Mathematics	38
2.5		Problems	40

3 Fundamentals of Measure Theory — 41
- 3.1 Measuring History — 41
- 3.2 Measure in an Algebra of Sets — 44
- 3.3 The Riemann Integral — 47
- 3.4 The Lebesgue Integral — 54
  - 3.4.1 The Lebesgue Measure — 57
  - 3.4.2 Concept of the Lebesgue Integral — 61
  - 3.4.3 Properties of the Lebesgue Integral — 62
- 3.5 Henri Lebesgue — 66
- 3.6 Problems — 67

4 Generalized Functions — 69
- 4.1 A Note on Generalized Functions — 69
- 4.2 The Unit Step Function — 70
  - 4.2.1 Properties of the Unit Step Function — 71
- 4.3 The Signum Function — 72
- 4.4 The Gate Function — 73
- 4.5 The Impulse Function — 73
  - 4.5.1 The Functional — 79
  - 4.5.2 Properties of the Impulse Function — 81
  - 4.5.3 Composite Function with the Impulse — 82
- 4.6 Doublet Generalized Function — 85
- 4.7 The Ramp Function — 86
- 4.8 The Exponential Function — 87
- 4.9 Discrete Functions — 89
  - 4.9.1 Discrete Unit Step Function — 90
  - 4.9.2 Discrete Impulse Function — 90
  - 4.9.3 Discrete Ramp Function — 91
- 4.10 Paul Dirac — 94
- 4.11 Problems — 95

5 Probability Theory — 97
- 5.1 Reasoning in Games of Chance 97
- 5.2 Measurable Space 99
  - 5.2.1 Probability Measure 100
  - 5.2.2 Probability Measure with the Riemann Integral ... 100
  - 5.2.3 Probability Measure with the Lebesgue Integral ... 101
- 5.3 The Axioms of Probability 105
- 5.4 Axioms of the Expectation Operator 108
- 5.5 Bayes' Theorem 109
- 5.6 Andrei Kolmogorov 113
- 5.7 Problems 114

6 Random Variables — 117
- 6.1 The Concept of a Random Variable 117
  - 6.1.1 Algebra Generated by a Random Variable 119
  - 6.1.2 Lebesgue Measure and Probability 119
- 6.2 Cumulative Distribution Function 120
  - 6.2.1 Change of Variable Theorem 126
- 6.3 Moments of a Random Variable 127
  - 6.3.1 Properties Associated to the Expected Value 128
  - 6.3.2 Definition of the Most Important Moments 129
- 6.4 Functions of Random Variables 132
  - 6.4.1 General Formula for Transformation 138
- 6.5 Discrete Distributions 144
- 6.6 Characteristic Function 149
- 6.7 Conditional Distribution 151
- 6.8 Useful Distributions and Applications 156
- 6.9 Carl Friedrich Gauss 164
- 6.10 Problems 165

7 Joint Random Variables — 167
- 7.1 An Extension of the Concept of Random Variable 167
- 7.2 Properties of Probability Distributions 171
- 7.3 Moments in Two Dimensions 172
- 7.4 Conditional Moments 177
- 7.5 Two-Dimensional Characteristic Function 178
  - 7.5.1 Sum of Random Variables 180
- 7.6 Function of Joint Random Variables 181
- 7.7 Transformation of Random Vectors 186

x   Contents

	7.8	Complex Random Variables	188
	7.9	Félix Borel	190
	7.10	Problems	191

8	**Probability Fundamental Inequalities**		**193**
	8.1	Historical Notes	193
	8.2	Tchebychev's Inequality	195
	8.3	Markov's Inequality	197
	8.4	Bienaymé's Inequality	198
	8.5	Jensen's Inequality	198
	8.6	Chernoff's Inequality	199
	8.7	Kolmogorov's Inequality	201
	8.8	Schwarz' Inequality	202
	8.9	Hölder's Inequality	203
	8.10	Lyapunov's Inequality	204
	8.11	Minkowsky's Inequality	204
	8.12	Fatou's Lemma	205
	8.13	About Arguments and Proofs	207
	8.14	Problems	208

9	**Convergence and the Law of Large Numbers**			**211**
	9.1	Forms of Convergence in Probability Theory		211
	9.2	Types of Convergence		211
		9.2.1	Convergence in Probability	212
		9.2.2	Almost Sure Convergence	212
		9.2.3	Sure Convergence	213
		9.2.4	Convergence in Distribution	214
		9.2.5	Convergence in Mean of Order r	214
		9.2.6	Convergence in Mean	214
		9.2.7	Convergence in Mean Square	215
		9.2.8	Convergence in Measure	216
	9.3	Relationships Between the Types of Convergence		216
	9.4	Weak Law of Large Numbers		217
	9.5	Strong Law of Large Numbers		219
	9.6	Central Limit Theorem		222
		9.6.1	Demonstration of the Theorem	222
		9.6.2	Central Limit Theorem for Products	225
	9.7	Pierre-Simon Laplace		226
	9.8	Problems		227

A	**Formulas and Important Inequalities**	**231**
B	**Fourier Transform**	**239**
	B.1 Table of Fourier Transforms	239
C	**Commented Bibliography**	**247**

Bibliography   **259**

Index   **265**

About the Authors   **275**

Preface

Probability theory is a classic topic in the educational market, which evolved from the amalgamation of different areas of mathematics, such as set theory, developed by Georg Cantor, and measure theory, fostered by Henri Lebesgue, until it received a full axiomatic treatment by Andrei Kolmogorov, in the previous century. The subject studies the analysis methods that are usual to deal with random phenomena.

The theory is fundamental to most areas of knowledge, including engineering, mathematics, physics, sciences, economics, social sciences, finance, biology, and health sciences. It is an important part of the syllabus for most courses in mathematics and engineering.

For electrical engineering courses, it is a pre-requisite to most disciplines, including stochastic processes, communication systems, source coding, modulation systems, transmission techniques, channel coding, signal analysis, information theory, stochastic control, estimation, and digital signal processing. This book is designed for a course in modern probability based on measure theory.

A study published by the important Institute of Electrical and Electronics Engineers (IEEE) revealed that the companies, enterprises, and industries search for professionals with a solid background in mathematics and sciences, instead of a specialized professional from the previous century. The employment market in this area is in demand of information technology professionals and engineers who could afford to change and learn, as the market changes. This means that the market needs professionals who can model and design using mathematical tools.

Along the years, few books were published covering all the subjects needed to understand the very fundamental concepts of probability theory, from a set theoretical perspective that uses concepts of measure theory. Most books that deal with the subject are destined to very specific audiences. The vast majority of the books introduce probability using the relative frequency approach, which is appealing to statistics but completely outdated for probability theory. Some books still use the classical approach, which is useful to introduce applications but which is being abandoned by new authors who prefer a more fundamental and modern approach.

The more mathematically oriented books are seldom used by people with engineering, economics, or statistical background because the authors are more interested in theorems and related conditions than in fundamental concepts and applications. The books written for engineers usually lack the required rigor or skip some important points in favor of simplicity and conciseness.

The idea is to present a seamless connection between the more abstract advanced set theory, the fundamental concepts from measure theory, and integration to introduce the axiomatic theory of probability, filling in the gaps from previous books and leading to an interesting, robust, and, hopefully, self-contained exposition of the theory.

The book presents an account of the historical evolution of probability theory as a mathematical discipline. Each chapter presents a short biography of the important scientists who helped develop the related subject. Chapter 1 deals with the advanced concepts of set theory, which include algebra and cardinality. Chapter 2 presents the fundamental concepts of relation and function. The fundamentals of measure theory are discussed in Chapter 3. Generalized functions, including unit step, impulse, ramp, and their discrete counterparts, are discussed in Chapter 4.

The axioms of probability, as proposed by Andrei Kolmogorov, are presented in Chapter 5. Real random variables in one dimension are the subject of Chapter 6. Real and complex random variables in two dimensions, including moments and characteristic function, are covered in Chapter 7. Some fundamental inequalities related to the probability theory are discussed in Chapter 8. Chapter 9 presents the main convergence theorems for random variables, the law of large numbers, and the central limit theorem, both for sum and product.

Appendix A presents useful formulas, definite and indefinite integrals, series expansions, summations, and important identities and inequalities. Appendix B presents the Fourier transform, some of its properties, and a collection of important transforms. Appendix C includes a commented review of several books on sets, measure, and probability theory, which will be helpful to the reader. Appendix D presents information about the authors. The book also has an extended bibliography and an index to help the readers find the required information.

Marcelo S. Alencar, Raphael T. Alencar

Acknowledgements

The publication of this book is the result of the many years of the author's work at the Federal University of Rio Grande do Norte (UFRN), Natal, at the Senai Cimatec University Center, Salvador, at the Federal University of Bahia (UFBA), Salvador, at the Federal University of Campina Grande (UFCG), Campina Grande, at the University for the Development of the State of Santa Catarina (UDESC), Joinville, at the Raydiall Automotive, Voiron, France, and at the Institute of Advanced Studies in Communications (Iecom).

It also ensues from the privilege of cooperating with several companies, firms, and institutions, through time, among which are the Brazilian Telecommunications Company (Embratel), the Technical-Scientific Association Ernesto Luiz de Oliveira Junior (Atecel), the Brazilian Company of Mail and Telegraphs (Correios do Brasil), the Telecommunications of Rio Grande do Norte (Telern), the Hydroelectric Company of the São Franciso (Chesf), and the Telecommunications of Paraíba (Telpa).

This cooperation also involved the Telecommunications of Pernambuco (Telpe), the Tele Nordeste Celular (TIM), The National Agency of Telecommunications (Anatel), Siemens of Brazil S.A., Oi, Brazilian Telecommunications S.A. (Telebrás), Alpargatas S/A, and Licks Attorneys. The experience this text mirrors to the readers is the fruit of this long-date cooperation. The authors also thank the complete revision of the text, done by Junko Nakajima.

The understanding and affection of Silvana, Thiago, Raissa, Raphael, Janaina, Marcella, Vicente, and Cora, who forgave the long periods of absence on account of the academic work, allowed the authors to develop this book, based on articles published in journals, magazines, and conferences, in addition to articles written by the first author for the column Difusão, published by the Jornal do Commercio, Recife, Brazil.

Marcelo S. Alencar, Raphael T. Alencar

List of Figures

Figure 1.1	A Venn diagram that represents two intersecting sets.	4
Figure 1.2	A Venn diagram representing disjoint sets.	4
Figure 1.3	Increasing sequence of sets.	8
Figure 1.4	Decreasing sequence of sets.	8
Figure 1.5	Mapping of the real set to the interval $[0, 1]$.	18
Figure 1.6	A portrait of Georg Ferdinand Ludwig Philipp Cantor. Adapted from: Public Domain, https://commons.wikimedia.org/w/index.php?curid=74820875	21
Figure 2.1	Relation in graph.	25
Figure 2.2	Relation expressed as a mapping, or Venn diagram.	25
Figure 2.3	A function is a subset of a relation.	28
Figure 2.4	Venn diagram for an injective function.	28
Figure 2.5	Venn diagram for a many to one function.	29
Figure 2.6	Venn diagram for a surjective function.	29
Figure 2.7	Venn diagram for a bijective function.	29
Figure 2.8	A function associates the element $a \in A$, in the domain, with the element $b \in B$, in the co-domain.	32
Figure 2.9	A function maps the set $C \subset A$ to its image $D \subset B$.	32
Figure 2.10	The inverse function associates the element $b \in B$, of the range, with the element $a \in A$.	33
Figure 2.11	Example of a fuzzy indicator function.	35
Figure 2.12	The inverse function of the union of sets A and B is the union of the inverse functions of A and B.	36
Figure 2.13	The union function of the sets A and B is the union of the functions of A and B.	37
Figure 2.14	A portrait of Johan Maurits van Nassau-Siegen. Adapted from: Public Domain, https://commons.wikimedia.org/w/index.php?curid=2145765	38
Figure 3.1	How to compute the Riemann integral.	48

Figure 3.2	Computation of the Riemann integral, for the identity function.	49
Figure 3.3	Sketch of the Riemann integral, upper limit.	52
Figure 3.4	The lower limit for the Cauchy integral.	53
Figure 3.5	An illustration of the Stieltjes integral.	55
Figure 3.6	An illustration of the Lebesgue integral.	56
Figure 3.7	A sketch of the Lebesgue method.	56
Figure 3.8	The symmetric difference set $A \triangle U$.	59
Figure 3.9	Quantization process.	64
Figure 3.10	A portrait of Henri Léon Lebesgue. Adapted from: Public Domain, https://commons.wikimedia.org/w/index.php?curid=336482	67
Figure 4.1	A unit step function.	71
Figure 4.2	A square wave produced from a cosine wave, using the unit step function.	72
Figure 4.3	The signum function.	73
Figure 4.4	The Gate function.	74
Figure 4.5	The impulse function.	74
Figure 4.6	The limiting definition of the impulse function.	75
Figure 4.7	An infinite series of impulses.	77
Figure 4.8	The impulse function shifted to τ to sample the function $f(x)$.	78
Figure 4.9	Periodic series of alternating impulses.	78
Figure 4.10	The function $\mathrm{u}[f(x)]$.	83
Figure 4.11	Parabola that crosses the abscissa at two points.	84
Figure 4.12	Doublet generalized function.	85
Figure 4.13	Ramp function.	87
Figure 4.14	The exponential function.	88
Figure 4.15	The derivative of the limited exponential function.	89
Figure 4.16	The discrete version of the unit step function.	90
Figure 4.17	The discrete version of the impulse function.	91
Figure 4.18	The discrete version of the ramp function.	92
Figure 4.19	The discrete version of the window function, for $n = 2$.	93
Figure 4.20	The discrete version of the exponential function.	93
Figure 4.21	The discrete version of the complex exponential function.	94

List of Figures   xix

Figure 4.22	A portrait of Paul Adrien Maurice Dirac. Adapted from: Public Domain, http://nobelprize.org/nobel prizes/physics/laureates/1933/dirac.html	95
Figure 5.1	Measure of the Cantor set.	104
Figure 5.2	Venn diagram of the intersection of sets A and B.	110
Figure 5.3	Partition of set B by a family of sets $\{A_i\}$.	111
Figure 5.4	Partition of a set.	112
Figure 5.5	Andrei Kolmogorov was one of the more important scientists of the 20th Century (Source: Konrad Jacobs/Wikimedia Commons).	113
Figure 6.1	Mapping of the events in the random variable domain.	118
Figure 6.2	Cumulative distribution function.	121
Figure 6.3	The probability density function.	122
Figure 6.4	Exponential probability density function, with parameter $a = 1$.	123
Figure 6.5	The area under the probability density function, from a to b is the difference between the respective cumulative distributions.	124
Figure 6.6	The probability measurement using the cumulative distribution function.	124
Figure 6.7	Laplace probability density function, with parameters $a = 1$ and $b = 0$.	125
Figure 6.8	Laplace cumulative distribution function, with parameters $a = 1$ and $b = 0$.	126
Figure 6.9	Gaussian probability density function, for three different values of the mean, μ, and standard variation, σ_X.	131
Figure 6.10	Gaussian cumulative distribution function, for three different values of the mean, μ, and standard deviation, σ_X.	132
Figure 6.11	Application of transformation of pdf to the solution of a circuit, described by the equation $Y = g(X)$.	133
Figure 6.12	The bijective mapping implies equivalent areas.	135
Figure 6.13	Illustration for the pdf transformation caused by a linear amplifier.	137
Figure 6.14	Chi-square probability density function, for three different values of the parameter n.	140

xx *List of Figures*

Figure 6.15	Chi-square cumulative distribution function, for three different values of the parameter n.	141
Figure 6.16	Result for the distribution of a Gaussian signal that goes through a full-wave rectification.	142
Figure 6.17	Uniform distribution.	142
Figure 6.18	The distribution function obtained from the mapping $y = V \cos \phi$.	143
Figure 6.19	Bernoulli probability density function.	145
Figure 6.20	Bernoulli cumulative distribution function.	145
Figure 6.21	Geometric probability density function.	146
Figure 6.22	Geometric cumulative distribution function.	147
Figure 6.23	Poisson probability density function.	148
Figure 6.24	Poisson cumulative distribution function.	149
Figure 6.25	Line segments used in conditional probability.	152
Figure 6.26	Line segments used in the example.	152
Figure 6.27	Conditional pdf and CDF.	153
Figure 6.28	Line segments used to calculate the conditional probability.	154
Figure 6.29	Conditional probability distributions.	155
Figure 6.30	Additive noise channel, $Y = A + X$.	155
Figure 6.31	Rayleigh probability density function.	157
Figure 6.32	Rayleigh cumulative distribution function.	158
Figure 6.33	Lognormal pdf.	160
Figure 6.34	Lognormal CDF.	161
Figure 6.35	A portrait of Johann Carl Friedrich Gauss. Adapted from: Public Domain, https://commons.wikimedia.org/w/index.php?curid=6886354	164
Figure 7.1	Region $\{-\infty < X \leq x, -\infty < Y \leq y\}$.	168
Figure 7.2	Region $\{-\infty < X \leq x_1, y_1 < Y \leq y_2\}$.	169
Figure 7.3	Region $\{x_1 < X \leq x_2, y_1 \leq Y \leq y_2\} = R_2$.	169
Figure 7.4	Region defined by a zero correlation coefficient.	173
Figure 7.5	Region defined by a positive correlation coefficient.	174
Figure 7.6	Region defined by a negative correlation coefficient.	175
Figure 7.7	Generic regions used to analyze the transformation.	181
Figure 7.8	Region $\{x < X \leq x + \mathrm{d}x, y < Y \leq y + \mathrm{d}y\}$.	182
Figure 7.9	Differential region for the output joint variables.	183
Figure 7.10	Linearized output differential region.	183
Figure 7.11	Representation of a complex random variable in the complex plane.	188

Figure 7.12	A portrait of Félix Edouard Juston Émile Borel. Adapted from: Public Domain,	190
Figure 8.1	Timon of Phlius, Ancient Greek skeptic philosopher. Source: Thomas Stanley, 1655, The History of Philosophy, Public Domain, https://commons.wikimedia.org/w/index.php?curid=8722615	208
Figure 9.1	A portrait of Pierre-Simon, marquis de Laplace. Adapted from: James Posselwhite, www.britannica.com, Public Domain, https://commons.wikimedia.org/w/index.php?curid=11128070	226

List of Tables

Table 2.1 Relation in tabular form. 25

List of Abbreviations

AC	Alternate current
AWGN	Additive white Gaussian noise
CDF	Cumulative distribution function
DC	Direct current
FFT	Fast Fourier transform
i.i.d.	Independent and identically distributed
IEEE	Institute of Electrical and Electronics Engineers
IFT	Inverse Fourier transform
LS	Lower sum
pdf	Probability density function
RMS	Root mean square
SNR	Signal-to-noise ratio
UDESC	University for the Development of the State of Santa Catarina
UFBA	Federal University of Bahia
UFCG	Federal University of Campina Grande
UFRN	Federal University of Rio Grande do Norte
US	Upper sum

1

Advanced Set Theory

"The essence of mathematics lies in its freedom."
Georg Cantor

1.1 Set Theory

Probability is the science of random phenomena, in a sense that it studies the properties of events that depend mainly on the concept of randomness. It is a branch of mathematics, built using the axiomatic method. Some abstract concepts are accepted as true, such as the notion of a set, and relations are constructed between those concepts that justify the theory. Then, based on logical deduction only, certain theorems are obtained and the theory is applied to real life (Parzen, 1979).

The modern theory of sets was developed by the Prussian mathematician Georg Ferdinand Ludwig Philipp Cantor (1845–1918), in the end of the 19th century. After concluding his doctorate under the supervision of the Polish mathematician, Leopold Kronecker (1823–1891), at the Humboldt University, in Germany, Cantor established the mathematical and logical basis for the theory, and demonstrated several important results.

The mathematical concept of set cardinality and the notion of denumerable and non-denumerable sets, defined by Cantor, were fundamental to the modern theory of sets. Cantor was born in Saint Petersbourg, Russia, but lived most of his academic life in the city of Halle, Germany, which is famous for the University of Halle and for the production of rock salt (Boyer, 1974).

The basic ideas of universal set, empty set, set partition, discrete systems, continuous systems, and infinity are as old as humanity itself. Over the years, philosophers and mathematicians had tried to characterize the infinite, with little success. In 1872, Julius Wilhelm Richard Dedekind (1831–1916), a German mathematician, discovered the universal property of infinite sets. A set is called infinite when it is similar to a part of itself; in the contrary case, the set is finite (Boyer, 1974).

At the same time he received his habilitation, Cantor was appointed to a position at the University of Halle, in the medieval city of Halle in Germany. There, Cantor investigated the properties of infinite sets and concluded that the infinite sets did not always have the same cardinality. This observation led to the concept of cardinal numbers, to establish a hierarchy of infinite sets in accordance with their respective powers.

In 1873, Cantor proved that the rational numbers were countable, that is, they may be placed in one-one correspondence with the natural numbers. He also showed that the algebraic numbers, that is, the numbers which are roots of polynomial equations with integer coefficients, were countable. His proof that the real numbers were not countable was published in a paper, in 1874.

Cantor's ideas established set theory as a complete subject. As a consequence of his results on transfinite arithmetic, considered advanced for his own time, Cantor suffered attacks from mathematicians like Leopold Kronecker, who also barred him from a position at the University of Berlin.

Between 1879 and 1884, Cantor published six papers in the periodical *Mathematische Annalen*, providing a basic introduction to set theory. In 1884, Cantor had the first recorded attack of depression, which was followed by many others. He died of a heart attack in an institution for mental health treatment, in Halle, following his attempts to apply his theory to justify religious paradigms in scientific events.

In fact, a probabilistic event is considered a set of descriptions. This implies that, if one decides that a certain event A has occurred it means that the result of a certain random situation has as description an element of the set A, which is usually called an outcome of the theoretical experiment (Parzen, 1979).

1.2 Basic Set Theory

The notion of a set is axiomatic, because it does not admit a definition that does not resort to the original notion of a set. On the other hand, the mathematical concept of a set is fundamental in Mathematics, and is used to build important concepts, such as relation, Cartesian product and function. It is also the basis for the contemporary measure theory, developed by Henry Léon Lebesgue (1875–1941).

Advanced set theory is based on few independent and fundamental axioms: Axiom of Extension, Axiom of Specification, Peano's Axioms and

1.2 Basic Set Theory

Axiom of Choice, besides Zorn's Lemma and Schröder-Bernstein's Theorem (Halmos, 1960).

The objective of this section is to present the theory of sets in an informal manner, just quoting those fundamental axioms, since this theory is used as a basis to establish a probability measure. There are more specific books on the subject, including (Braumann, 1987) (Gödel, 1979) (Ash, 1990). Some examples of common sets are given here.

- The binary set: $\mathbb{B} = \{0, 1\}$;
- The set of natural numbers, including zero: $\mathbb{N} = \{0, 1, 2, 3, \ldots\}$;
- The set of odd numbers: $\mathbb{O} = \{1, 3, 5, 7, 9, \ldots\}$;
- The set of integer numbers: $\mathbb{Z} = \{\ldots, -3, -2, -1, -2, 0, 1, 2, 3, \ldots\}$;
- The set of real numbers: $\mathbb{R} = (-\infty, \infty)$.
- The set of complex numbers: $\mathbb{C} = \{a + jb \mid a, b \in \mathbb{R}, j = \sqrt{-1}\}$.

Two relations are important in set theory: the belonging relation, denoted as $a \in A$, which implies a is an element of the set A, and the inclusion relation, $A \subset B$, which is read as "A is a subset of the set B", or B is a super set of the set A. Sets may also be specified by propositions. For example, the empty set can be written as $\emptyset = \{a \mid a \neq a\}$, that is, the set the elements of which are different from themselves.

It is important to differentiate between inclusion and belonging, which are conceptually different mathematical relations. For instance, inclusion is always reflexive, whereas it is not possible to say that belonging is ever reflexive. For example, the relation $A \subset A$ is always true. On the other hand, it is not possible, or reasonable, to say that $A \in A$ is ever true. Additionally, inclusion is transitive, but belonging is not (Halmos, 1960).

A universal set contains all other sets of interest. An example of a universal set is provided by the sample space in probability theory, usually denoted by S or Ω. It is interesting to note that the universal set does not contains all other sets, as is it erroneously defined, because this notion is a known paradox, attributed to Bertrand Arthur William Russel (1872–1970), an influential philosopher, mathematician and politician from Wales. An example of a universal set of interest is the set of integers $\mathbb{Z}$.

The empty set is that set which contains no element, and is usually denoted by $\emptyset$ or $\{\ \}$. It is implicit that the empty set is contained in any set, that is, that $\emptyset \subset A$, for any given set A. However, the empty set is not, in general, an element of other sets. The set of odd numbers that are also even is an example of an empty set.

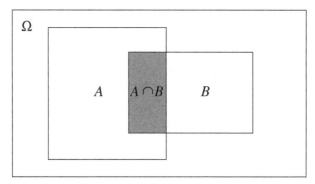

Figure 1.1 A Venn diagram that represents two intersecting sets.

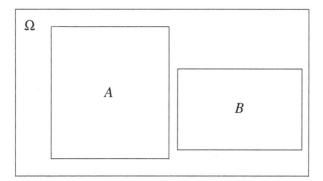

Figure 1.2 A Venn diagram representing disjoint sets.

The Venn diagram was created by John Venn (1834–1923), a British priest and mathematician. It is illustrated in Figure 1.1 and is a common way to represent and deal with sets.

Sets are disjoint if they have no element in common, as illustrated in Figure 1.2. Thus, for example, the set of even natural numbers and the set of odd natural numbers are disjoint.

1.3 The Axioms of Set Theory

The following axiom system was formulated in first-order logic, with equality, and with only one binary relation symbol ∈ to indicate membership. It is known as the Zermelo–Fraenkel axiom system, named after the German mathematicians Ernst Friedrich Zermelo (1871–1953) and Abraham Abraham Halevi Fraenkel (1891-1965) and includes the Axiom of Choice (Bagaria, 2014).

- Null Set – There is a set denoted by $\emptyset$ and called the empty set, which has no elements.
- Extension – Two sets A and B that have the same elements are equal.
- Pair – Given any two sets A and B, there is a set, denoted by $\{A, B\}$, which contains A and B as its only elements. In particular, the set $\{A\}$ has A as its only element.
- Power Set – For every set A there exists a set, denoted by 2^A, called the power set of A, whose elements are all the subsets of A.
- Union – For every set A, there is a set, denoted by $\cup A$, called the union of A, whose elements are all the elements of A.
- Infinity – There is an infinite set. In particular, there exists a set $\mathcal{A}$ that contains $\emptyset$, such that if $A \in \mathcal{A}$, then $\cup\{A, \{A\}\} \in \mathcal{A}$.
- Separation – For every set A and a given property, there is a set containing exactly the elements of A that have the property.
- Replacement – For every given definable function whose domain is a set A, there is a set whose elements are all the values of the function.
- Foundation – Every non-empty set A contains an ϵ-minimal element, that is, an element such that no element of A belongs to it.
- Choice – For every set A of pairwise-disjoint non-empty sets, there exists a set that contains exactly one element from each set in A. This is a controversial axiom, but an important principle, and it is very useful and widely used in Mathematics.

This last axiom can be stated as: the Cartesian product of a non-empty family of non-empty sets is non-empty, and implies that if $\{A_i\}$ is a family of non-empty sets indexed by a non-empty set I, then there is a family $\{a_i\}, i \in I$, such that $a_i \in A_i$ for each $i \in I$ (Halmos, 1960).

1.4 Operations on Sets

Sets can be combined in several ways. It is possible to operate on sets to produce new sets, or families of sets. The basic set operations are the complement, the union, the intersection, the subtraction and the symmetric difference. For instance, the complement of a union is the intersection of the complements and the complement of an intersection is the union of the complements (Hausdorff, 2005).

- The operation $\overline{A}$ represents the complement of A with respect to the sample space Ω;

- The union of two sets is composed of elements that belong to A or to B, and is written as $A \cup B$;
- The intersection of two sets is composed of elements that belong to A and to B, and is written as $A \cap B$;
- The subtraction of sets, denoted by $C = A - B$, gives as a result the set the elements of which belong to A and do not belong to B.
  Note: If B is completely contained in A, then $A - B = A \cap \overline{B}$;
- The symmetric difference is defined as the set of elements that belong to A and to B, but do not belong to $(A \cap B)$. It is commonly written as $A \triangle B = A \cup B - A \cap B$.

The generalization of these concepts to families of sets, as for example, $\cup_{i=1}^{N} A_i$ and $\cap_{i=1}^{N} A_i$, is immediate. The following properties are usually employed as axioms in developing the theory of sets (Lipschutz, 1968).

- **Idempotent**
  $A \cup A = A, \qquad A \cap A = A$
- **Associative**
  $(A \cup B) \cup C = A \cup (B \cup C), \qquad (A \cap B) \cap C = A \cap (B \cap C)$
- **Commutative**
  $A \cup B = B \cup A, \qquad A \cap B = B \cap A$
- **Distributive**
  $A \cup (B \cap C) = (A \cup B) \cap (A \cup C),$
  $A \cap (B \cup C) = (A \cap B) \cup (A \cap C)$
- **Identity**
  $A \cup \emptyset = A, \qquad A \cap \Omega = A$
  $A \cup \Omega = \Omega, \qquad A \cap \emptyset = \emptyset$
- **Complementary**
  $A \cup \overline{A} = \Omega, \qquad A \cap \overline{A} = \emptyset \qquad \overline{(\overline{A})} = A$
  $\overline{\Omega} = \emptyset, \qquad \overline{\emptyset} = \Omega$
- **De Morgan laws**, introduced by the British logician and mathematician Augustus De Morgan (1806–1871), born in Madurai, India.
  $\overline{A \cup B} = \overline{A} \cap \overline{B}, \qquad \overline{A \cap B} = \overline{A} \cup \overline{B}$

Aside from the mentioned properties, one can also define the *supremum* of a set $A = \{x_1, x_2, \dots\}$ of real numbers, as its smallest limit superior (Taylor, 1985),

$$\liminf_{n \to \infty} x_n = \sup \inf_{k \geq 1} \{x_k, x_{k+1}, \dots\}. \qquad (1.1)$$

and the *infimum* of A as its greatest limit inferior,

$$\limsup_{n\to\infty} x_n = \inf \sup_{k\geq 1}\{x_k, x_{k+1}, \ldots\}. \quad (1.2)$$

It is possible to demonstrate that

$$\liminf_{n\to\infty} x_n \leq \limsup_{n\to\infty} x_n. \quad (1.3)$$

Example: Consider that $x_n = 1 + \frac{1}{n}$, if n is odd, and $x_n = -\frac{1}{n}$, if n is even. Then,

$$\liminf_{n\to\infty} x_n = -\infty$$

and

$$\limsup_{n\to\infty} x_n = 1. \blacksquare$$

Example: Consider the set $\{x_n\}$, whose elements are defined as follows,

$$x_n = 2 + \frac{(-1)^n}{n}, \text{ if } n = 1, 4, 7, 10, \ldots,$$

$$x_n = 1 + \frac{1-(-1)^n}{n}, \text{ if } n = 2, 5, 8, 11, \ldots,$$

$$x_n = 0, \text{ if } n = 3, 6, 9, 12, \ldots.$$

Then,

$$\liminf_{n\to\infty} x_n = 0$$

and

$$\limsup_{n\to\infty} x_n = 2. \blacksquare$$

1.5 Families of Sets

In set theory, a collection of subsets of a given set is called a family of subsets. The concept of family is important to characterize finite or infinite combinations of sets. The increasing sequence of sets, such that

$$\lim_{i\to\infty} \cup A_i = A, \quad (1.4)$$

is one of the most useful families of sets. Figure 1.3 illustrates an increasing sequence of sets. This sequence is used in proofs of limits over sets.

8 Advanced Set Theory

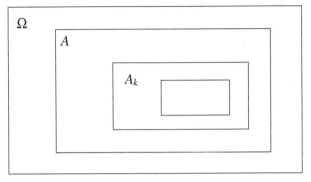

Figure 1.3 Increasing sequence of sets.

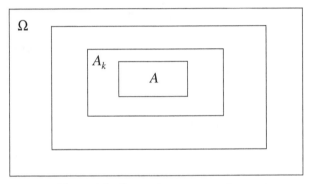

Figure 1.4 Decreasing sequence of sets.

The decreasing sequence of sets is defined in a similar manner, as follows:

$$\lim_{i \to \infty} \cap A_i = A, \qquad (1.5)$$

and is also used in proofs of limits over sets. Figure 1.4 illustrates a decreasing sequence of sets.

The limit superior of a sequence can be defined as a limiting bound on the sequence. In this regard, the limit superior of a sequence $\{A_n, n \geq 1\}$, that is, the set of elements of Ω that belong to an infinite number of A_n, is defined as (Magalhães, 2006),

$$\limsup A_n = \bigcap_{n=1}^{\infty} \bigcup_{k=n}^{\infty} A_k. \qquad (1.6)$$

Note that ω belongs to Equation (1.6) if and only if for each n there is some $k \geq n$ for which $\omega \in A_k$. In other words, ω is in (1.6) if and only if it belongs to infinitely many sets $A_k n$ (Billingsley, 1995).

The limit inferior of a sequence is a limiting bound on the sequence. This means that the limit inferior of a sequence of sets $\{A_n, n \geq 1\}$, that represents the set of elements Ω that belong to all A_n from a given n, is defined as

$$\liminf A_n = \bigcup_{n=1}^{\infty} \bigcap_{k=n}^{\infty} A_k. \qquad (1.7)$$

Note that ω is in Equation (1.7) if and only if there is some n such that $\omega \in A_k$ for all $k \geq n$. In other words, ω is in Equation (1.6) if and only if it belongs to all, except a finite set of $A_k n$.

It can be verified that

$$\bigcup_{k=n}^{\infty} A_k \downarrow \limsup A_n,$$

$$\bigcap_{k=n}^{\infty} A_k \uparrow \liminf A_n.$$

and, for all m and n

$$\bigcap_{k=n}^{\infty} A_k \subset \bigcup_{k=n}^{\infty} A_k,$$

because, for $i \geq \max\{m, n\}$, A_i contains the first of those sets and is contained in the second. Taking the union in m and the intersection in n it is possible to show that Equation (1.7) is a subset of Equation (1.6).

When the limits superior and inferior of a sequence of sets $\{A_n, n \geq 1\}$ coincide, it is said that the sequence has a limit, that is,

$$\limsup A_n = \liminf A_n = \lim_{n \to \infty} A_n. \qquad (1.8)$$

The following properties can be verified (Ash, 1972):

- The complement of the limit superior of a sequence of sets is the limit inferior of the complement of that sequence,

$$\overline{(\limsup A_n)} = \liminf \overline{A_n}, \qquad (1.9)$$

- The complement of the limit inferior of a sequence of sets is the limit superior of the complement of that sequence,

$$\overline{(\liminf A_n)} = \limsup \overline{A_n}, \qquad (1.10)$$

- The limit inferior of a sequence of sets is contained in the limit superior of that sequence,

$$\liminf A_n \subset \limsup A_n. \qquad (1.11)$$

1.5.1 Indexing of Sets

The Cartesian product, popularized by the French mathematician René Descartes (1596–1650), is a useful tool to express the idea of indexing of sets. Indexing expands the possibilities for the use of sets, and allows the generation of new entities, such as vectors and signals.

Example: Consider $A_i = \{-1, 1\}$. Starting from this set it is possible to construct an indexed sequence of sets by defining its indexing: $\{A_{i \in I}\}$, $I = \{0, \cdots, 7\}$. This family of indexed sets A_i constitutes a finite discrete sequence, that is, a vector. ∎

Example: Again, let $A_i = \{-1, 1\}$, but now use $I = \mathbb{Z}$, the set of positive and negative integers plus zero. It follows that $\{A_{i \in \mathbb{Z}}\}$, which represents an infinite series of -1's and 1's, that is, it represents a binary digital signal. For example, $\cdots - 1 - 111111 - 1 - 1 - 1 \cdots$. ∎

Example: Using the same set $A_i = \{-1, 1\}$, but now indexing over the set of real numbers, $\{A_{i \in I}\}$, in which $I = \mathbb{R}$, it is possible to form a signal which is discrete in amplitude but continuous in time. ∎

Example: Considering $A = \mathbb{R}$ and $I = \mathbb{R}$, the resulting set represents an analog signal, that is, a signal that is continuous in time and in amplitude. ∎

1.6 An Algebra of Sets

An algebra of sets defines the fundamental properties of sets, and provides systematic procedures to evaluate expressions, and perform calculations, involving certain operations and relations. The usual set-theoretic operations are union, intersection, and complementing, and the relations of set equality and set inclusion. For the construction of an algebra of sets or, equivalently, for the construction of a field over which operations involving sets make sense, a few properties have to be obeyed.

1.6 An Algebra of Sets

1. If $A \in \mathcal{F}$ then $\overline{A} \in \mathcal{F}$. A is the set containing desired results, or over which one wants to operate;

2. If $A \in \mathcal{F}$ and $B \in \mathcal{F}$, then $A \cup B \in \mathcal{F}$.

The properties guarantee the closure of the algebra with respect to finite operations over sets. It is noticed that the universal set Ω always belongs to the algebra, that is, $\Omega \in \mathcal{F}$, because $\Omega = A \cup \overline{A}$. The empty set also belongs to the algebra, that is, $\emptyset \in \mathcal{F}$, since $\emptyset = \overline{\Omega}$, follows by property 1.

Example: The family $\{\emptyset, \Omega\}$ complies with the above properties and therefore represents an algebra. In this case $\emptyset = \{\}$ and $\overline{\emptyset} = \Omega$. The union is also represented, as can be verified. ■

Example: Given the sets $\{C_H\}$ and $\{C_T\}$, representing the faces of a coin, respectively, if $\{C_H\} \in \mathcal{F}$ then $\{\overline{C_H}\} = \{C_T\} \in \mathcal{F}$. It follows that $\{C_H, C_T\} \in \mathcal{F} \Rightarrow \Omega \in \mathcal{F} \Rightarrow \emptyset \in \mathcal{F}$. This can be explained by the following argument. If there is a measure for heads then there must also be a measure for tails, if the algebra is to be properly defined. Whenever a probability is assigned to an event then a probability must also be assigned to the complementary event. ■

The cardinality of a finite set is defined as the number of elements belonging to this set. Sets with an infinite number of elements are said to have the same cardinality if they are equivalent, that is, $A \sim B$ if $\sharp A = \sharp B$. Some examples of sets and their respective cardinals are presented next.

- $I = \{1, \cdots, K\} \Rightarrow C_I = K$;

- $\mathbb{N} = \{0, 1, \cdots\} \Rightarrow C_N = \aleph_0$

- $\mathbb{Z} = \{\cdots, -3, -2, -1, 0, 1, 2, 3, \cdots\} \Rightarrow C_Z = \aleph_0$

- $\mathbb{Q} = \{\cdots, -1/3, -1/4, 0, 1/4, 1/3, 1/2, \cdots\} \Rightarrow C_Q = \aleph_0$

- $\mathbb{R} = (-\infty, \infty) \Rightarrow C_R = \aleph$

For the above examples the following relations are verified: $C_\mathbb{R} > C_\mathbb{Q} = C_\mathbb{Z} = C_\mathbb{N} > C_I$. The notation for the first infinity, $\aleph_0$, was employed by Cantor. It uses the first letter of the Hebrew alphabet (Aleph) for the cardinality of the set of natural numbers,

Example: The cardinality of the power set, that is, the cardinality of the family of sets $\mathcal{F}$ consisting of all subsets of a given set I, $\mathcal{F} = 2^I$, is 2^{C_I}. For the previous example, $C_\mathcal{F} = 2^K$. ∎

1.7 The Borel Algebra

The Borel algebra was established by Félix Edouard Juston Émile Borel (1871–1956), a French mathematician and politician, who, together with René-Louis Baire and Henri Lebesgue, established the principles of Measure Theory and devised its applications to Probability Theory.

In 1893, at the age of 22, Borel presented his doctoral thesis, entitled *Sur quelques points de la théorie des fonctions*, supervised by the famous French mathematician Gaston Darboux (1842–1917). He was then appointed *maître de conférence* of the Université de Lille, where he worked for three years. He also worked at the *École Normale Supérieure*. He held a professorship at Sorbonne from 1909 to 1941.

The algebra is written as $\mathcal{B}$, and called σ-algebra. It is an extension of the algebra so far discussed to operate with limits at infinity. The following properties are required from a σ-algebra.

1. $A \in \mathcal{B} \Rightarrow \overline{A} \in \mathcal{B}$

2. $A_i \in \mathcal{B} \Rightarrow \bigcup_{i=1}^{\infty} A_i \in \mathcal{B}$

The properties guarantee the closure of the σ-algebra with respect to the enumerable operations over sets. They allow the definition of limits in the Borel field.

Example: Considering the above properties it can be verified that $A_1 \cap A_2 \cap A_3 \cdots \in \mathcal{B}$. In effect, it is sufficient to notice that

$$A \in \mathcal{B} \text{ and } B \in \mathcal{B} \Rightarrow A \cup B \in \mathcal{B},$$

and

$$\overline{A} \in \mathcal{B} \text{ and } \overline{B} \in \mathcal{B} \Rightarrow \overline{A} \cup \overline{B} \in \mathcal{B},$$

and finally

$$\overline{\overline{A} \cup \overline{B}} \in \mathcal{B} \Rightarrow A \cap B \in \mathcal{B}.$$

In summary, any combination of unions and intersections of sets belongs to the Borel algebra. In other words, operations of union or intersection of sets, or a combination of these operations, produce a set that belongs to the σ-algebra. ∎

1.8 Cardinality

The cardinality of a finite set is the number of elements of this set (Braumann, 1987) (Gödel, 1979). Cardinality is a measure that characterizes the sets, and represents and intuitive notion to count and compare them. For sets with infinite number of elements the concept of dimension is more abstract.

Georg Cantor developed a set of rules to treat these sets and showed that not all infinite sets have the same cardinality. This, of course, caused problems with mathematicians of the time, for whom the very notion of infinity was already strange – let alone dealing with distinct infinities.

1.8.1 Equivalence of Sets

Sets with the same number of elements have the same cardinality, only if they are equivalent, that is, $C_A = \sharp(A) = C_B = \sharp(B)$ if and only if $A \sim B$.

An equivalence relation in the set $\mathbb{U}$, that is, a subset R of $\mathbb{U} \times \mathbb{U}$, is that one that can be written $x \sim y$ to indicate that $(x, y) \in R$, with the following properties (Capiński and Kopp, 2005):

1. Reflexive – for all $x \in \mathbb{U}$, $x \sim x$;

2. Symmetric – $x \sim y$ implies $y \sim x$;

3. Transitive – $x \sim y$ and $y \sim z$ implies $x \sim z$.

Equivalent sets are those for which there is a bijective and surjective function $f : A \to B$, that connects each element of A to an element of B. A function with this property is also called invertible.

It is important to state that if A and B are sets, a function from A to B is a relation f such that the domain of f is B, and such that for each a in A there is a unique element b in B with $(a, b) \in f$. The symbol

$$f : A \to B \qquad (1.12)$$

is sometimes used as an abbreviation for the expression "f is a function on A into B." The set of all functions from A to B is a subset of the power set of the Cartesian product $A \times B$, sometimes denoted as B^A (Halmos, 1960).

1.8.2 Countable Sets

The sets equivalent to the set of natural numbers $\mathbb{N}$ are said to be enumerable, or countable. If it is not possible to find a one-to-one and surjective function between $\mathbb{N}$ and a given infinite set, this set is not enumerable or uncountable.

To check whether a given set A is enumerable, just find an indexing that uses the set of natural numbers for each element of $a_n \in A$. For example, just do

$$f(n) = a_n,$$

with $n \in \mathbb{N}$, to prove that A is countable.

The enumerable sets have the following properties (Kolmogorov and Formin, 1970):

- Every subset of a countable set is enumerable.
  Let A be countable, with elements $a_1, a_2, \ldots$, and let B be a subset of A. Among the elements $a_1, a_2, \ldots$, let $a_{n_1}, a_{n_2}, \ldots$ be those in set B. If the set of numbers $a_{n_1}, a_{n_2}, \ldots$ has a greater number, then B is finite. Otherwise, B is enumerable; just consider the correspondence $i \leftrightarrow a_{n_i}$.
- The union of a finite or countable number of countable sets $A_1, A_2, A_3, \ldots$ is enumerable.
  It can be assumed that $A_1, A_2, A_3, \ldots$ form an enumerable family of pairs of disjoint sets, each of them countable. If not, it is possible to form the family

  $$A_1, A_2 - A_1, A_3 - (A_1 - A_2), \ldots,$$

  which is countable by the previous property, and has the same union as the original set. To demonstrate the property, just write all elements of sets $A_1, A_2, A_3, \ldots$ in an infinite table

  $$\begin{array}{ccc} a_{11} & a_{12} & a_{13} \cdots \\ a_{21} & a_{22} & a_{23} \cdots \\ a_{31} & a_{32} & a_{33} \cdots \\ \vdots & \vdots & \vdots \end{array}$$

  in which the elements from set A_1 appear on the first line, the elements from set A_2 appear on the second line, and so on. Then, the elements are counted diagonally, a procedure created by Cantor, that is, the first chosen is a_{11}, the second is a_{12}, the third is a_{21}, as shown in the following table.

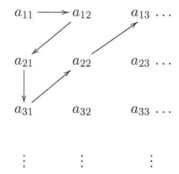

The procedure associates a single number with each element in each of the sets $A_1, A_2, A_3, \ldots$, thus establishing a correspondence between the union of the sets $A_1, A_2, A_3, \ldots$ and the set of natural numbers $\mathbb{N}$, that is, the union of the sets $\cup_{i \in \mathbb{N}} A_i$ is enumerable.

- Every infinite set has a countable subset.

Let B be an infinite set and a_1 be an element of B. Because it is infinite, B has an element a_2 distinct from a_1, an element a_3 distinct from a_1 and a_2, and so on. Continuing the process, which may not end due to lack of elements, since B is infinite, one obtains a countable set

$$A = \{a_1, a_2, \ldots, a_n, \ldots\}$$

of the set B. Note that countable sets are those with the lowest cardinality between all infinite sets.

Example: The set of odd numbers $\mathbb{O}$ can be put in an equivalence relation with the set of natural numbers $\mathbb{N}$, using the function $f : \mathbb{N} \to \mathbb{O}, f(a) = 2a + 1, a \in \mathbb{N}, b \in \mathbb{O}$. Therefore, the sets $\mathbb{O}$ and $\mathbb{N}$ have the same cardinality, even if $\mathbb{P}$ is a proper subset $\mathbb{N}$.∎

Example: Consider that the Cantor hotel receives a train with an infinite number of tourists. How is it possible to accommodate them in the hotel apartments if they are all fully booked?

The strategy is as follows. The manager asks each guest to move to the apartment whose number is twice his own, that is, the guest in the apartment number n moves to the apartment number $2n$. In this way, all odd rooms are available, and as the number of odd rooms is infinite, the hotel can accommodate the infinite number of tourists that have just arrived.∎

1.8.3 Uncountable Sets

Not every set is countable. The set of real numbers has the power of the continuum, that is, its cardinality $\aleph$ is greater than $\aleph_0$. In reality, the cardinality of the real set is $\aleph = 2^{\aleph_0}$. It is possible to directly prove that the set of real numbers is non denumerable, but it is instructive to prove that a subset of R is uncountable using Cantor's reasoning. Then, it suffices to consider that if a set has a non countable subset, it is also uncountable.

It can be proved that the interval $A = [0, 1]$ is not countable. To do so, consider a proof by contradiction, also known as *reductio ad absurdum*. Imagine that $A = \{x_1, x_2, x_3, \ldots\}$, that is, the elements of A can be placed in correspondence with the elements of the set of the natural numbers (Lipschutz, 1968). In this way, the elements of A can be written in the form of infinite decimals, as follows:

$$a_1 = 0.x_{11}x_{12}x_{13}\ldots x_{1n}\ldots$$
$$a_2 = 0.x_{21}x_{22}x_{23}\ldots x_{2n}\ldots$$
$$a_3 = 0.x_{31}x_{32}x_{33}\ldots x_{3n}\ldots$$
$$\vdots \quad \vdots$$
$$a_n = 0.x_{n1}x_{n2}x_{n3}\ldots x_{nn}\ldots \tag{1.13}$$

in which $x_{ij} \in \{0, 1, 2, \ldots, 9\}$ and each decimal contains an infinite number of non-zero digits. Numbers that can be written in two different ways must be placed as sequences with infinite nines, to avoid duplication.

It is possible to construct a real number

$$b = 0.y_1 y_2 y_3 \ldots y_n \ldots$$

that belongs to A in the following manner. Choose b such that $y_1 \neq x_{11}$, $y_2 \neq x_{22}$, $y_3 \neq x_{33} \ldots$ and $y_n \neq x_{nn}$. Note that $b \neq a_1$ because $y_1 \neq x_{11}$, $b \neq a_2$ because $y_2 \neq x_{22}$ and so on. That is, $b \neq a_n$ for all $n \in \mathbb{N}$. Thus, $b \notin A$, which contradicts the initial premise. Thus, the assumption that A is countable leads to a contradiction and therefore A is uncountable.

It can be concluded that the chance of finding rational numbers in the considered set is theoretically zero. To verify this, just think of the problem as the result of an experiment in which decimal digits are produced by a die with ten faces, numbered from zero to nine.

Note that rational numbers are formed by sequences that end in zero, those obtained by dividing multiple numbers, or by repeating decimals. In both cases there is an infinite repetition of digits. The probability of a die repeating itself infinitely in any experiment is effectively zero. Therefore, it is more likely to find only irrational numbers in the set.

Infinite sets have some interesting properties. For example, every infinite set is equivalent to one of its proper subsets, which is shown in the following.

As verified, any infinite set B contains a countable set. Consider that this set is
$$A = \{a_1, a_2, \ldots, a_n, \ldots\},$$
which can be partitioned into two countable subsets
$$A_1 = \{a_1, a_3, \ldots, a_{2n-1}, \ldots\} \text{ and } A_2 = \{a_2, a_4, \ldots, a_{2n}, \ldots\}.$$

It is possible to establish a one-to-one correspondence between the enumerable sets A and A_1; just do the following: $a_n \leftrightarrow a_{2n-1}$. This correspondence can be established between the sets $A \cup (B - A) = B$ and $A_1 \cup (B - A) = B - A_2$ simply by associating a itself to each element $a \in B - A$. However, $B - A_2$ is a proper subset of B.

The interval $[0, 1]$ has the power of the continuum and cardinality $\aleph$, just like the set of real numbers $\mathbb{R}$. To demonstrate this property, it is enough to find a one-to-one function defined on the set of reals.

Example: Consider the function $f : [0, 1] \mapsto [a, b]$ defined by
$$f(x) = a + (b - a)x.$$

This function is invertible and maps all elements from the closed interval $[0, 1]$ to the closed interval $[a, b]$, therefore $[0, 1] \sim [a, b]$, and there are countless elements in the range $[a, b]$. Making, for example, $a = -\pi/2$ and $b = \pi/2$, one can see that the function f maps the range $[0, 1]$ into $[-\pi/2, \pi/2]$. ∎

Example: The tangent function $g(x) = \tan x$ projects the interval $(-\pi/2, \pi/2)$ into the real line, $\mathbb{R}$, that is, it is possible to map $g : (-\pi/2, \pi/2) \mapsto (-\infty, \infty)$, since it is biunivocal and surjective in the considered domain. Therefore, any intervals on the line are equivalent to the set of reals and have power of the continuum. Particularly, $[0, 1] \sim \mathbb{R}$. ∎

Example: The function $f : (-\infty, \infty) \mapsto [0, 1]$, shown in Figure 1.5 and defined by
$$f(x) = \frac{1}{2}\left(1 + \frac{x}{1 + |x|}\right).$$
maps the real set to the interval $[0, 1]$. ∎

Additional examples of sets and their respective cardinals are presented in the following:

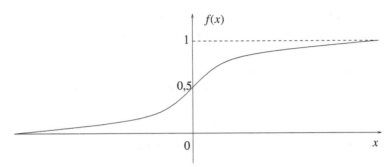

Figure 1.5 Mapping of the real set to the interval $[0, 1]$.

- Set of algebraic numbers, formed by the roots of polynomial functions, of degree higher or equal to one, with integer coefficients: $\mathbb{A} = \{\ldots, -\sqrt{3}, \ldots, \sqrt{-1}, \ldots, -\sqrt{2}, \ldots, 0, \ldots, 1/2, \ldots, \sqrt{10}, \ldots, \} \Rightarrow C_\mathbb{A} = \aleph_0$;
- Set of transcendental numbers, those that cannot be obtained by solving polynomial equations: $\mathbb{R} - \mathbb{A} = \{\ldots, -\ln 2, \ldots, e, \ldots, \pi, \ldots, e^{\sqrt{2}}, \ldots\} \Rightarrow C_{\mathbb{R}-\mathbb{A}} = \aleph$;
- Set of irrational numbers, which cannot be put in the form p/q, in which p and q are integers: $\mathbb{R} - \mathbb{Q} = \{\ldots, -\sqrt{3}, \ldots, \sqrt{2}, \ldots, \pi, \ldots\} \Rightarrow C_\mathbb{Q} = \aleph$;
- Set of complex numbers: $\mathbb{C} = \{a + jb \,:\, a, b \in \mathbb{R}, j = \sqrt{-1}\} \Rightarrow C_\mathbb{C} = \aleph$.

For the given examples, the following relations are verified: $C_\mathbb{C} = C_\mathbb{R} = C_{\mathbb{R}-\mathbb{A}} = C_{\mathbb{R}-\mathbb{Q}} > C_\mathbb{A} = C_\mathbb{Q} = C_\mathbb{Z} = C_\mathbb{N} > C_\mathbb{K}$. The notation $\aleph_0$ (*aleph zero*), for the cardinality of the set of natural numbers, was introduced by Cantor. The sets with power of the continuum have cardinality $\aleph$.

Example: The golden ratio $\varphi = \frac{1}{2}(1 + \sqrt{5}) \approx 1,618\ldots$ is an algebraic number, obtained by solving the polynomial equation $x^2 - x - 1 = 0$. This irrational number also can be obtained as a continued expansion of fractions

$$\varphi = \sqrt{1 + \sqrt{1 + \sqrt{\ldots}}}\blacksquare$$

Example: There is no function $f : \mathbb{R} \mapsto \mathbb{R}$, which is continuous everywhere, and that maps rational numbers to irrational ones, and vice versa.

1.8 Cardinality 19

Vito Volterra (1860–1940), a great Italian mathematician, was the first to demonstrate this impossibility (Dunham, 2005). Let $\mathbb{I} = \mathbb{R} - \mathbb{Q}$ be the set of irrational numbers, and suppose that $f(\mathbb{I}) \subseteq \mathbb{Q}$. Then, as the sets are disjoint, $f(\mathbb{R}) = f(\mathbb{Q}) \cup f(\mathbb{I}) \subseteq f(\mathbb{Q}) \cup \mathbb{Q}$, which is countable. But since f is continuous, $f(\mathbb{R})$ is connected. Therefore, $f(\mathbb{R})$ is countable and connected, that is, is a unitary set $\{x\}$ and f is constant.

Considering that no constant fulfills the required conditions, there is no continuous function such that $f(\mathbb{I}) \subseteq \mathbb{Q}$ and $f(\mathbb{Q}) \subseteq \mathbb{I}$. ∎

Example: The function $y = \tan x$, which is not continuous on the entire line, has the interesting property of mapping rational numbers to irrational numbers, and vice versa, that is, it works as a bridge between these complementary sets. ∎

Johann Heinrich Lambert (1728–1777), a Swiss mathematician based in Prussia, used this property to prove, in 1761, that π was an irrational number. Since $\tan \pi/4 = 1$, then $\pi/4$ must be irrational, hence π is irrational.

1.8.4 Cardinality Properties

There are some relationships between cardinal numbers, for operations with sets, which follow rules similar to those of arithmetic, as follows:

- The sum of cardinals is defined for disjoint sets. Let $\alpha = \sharp(A)$ and $\beta = \sharp(B)$ be the cardinal of the sets A and B. The sum is then $\alpha + \beta = \sharp(A \cup B)$.
- The multiplication of cardinals is defined for the Cartesian product of the sets A and B, given by $A \times B$, that is, $\alpha\beta = \sharp(A \times B)$. The Cartesian product is the set formed by all combinations of elements of A and B.
- The set B^A defines all possible functions, or mappings, of A into B. In this way, one defines the exponential operation for cardinal numbers like $\beta^\alpha = \sharp(B^A)$.
- The cardinality of the power set $\mathcal{F} = 2^A$, that is, the family formed by all subsets of a set A, equals 2^α. It turns out that the cardinality of the power set is always greater than that of the original set, so $\sharp(A) < \sharp(2^A)$ (Braumann, 1987).

The relationship between a set and its power set is given by the Cantor's Theorem. This theorem uses the inequality relationship between cardinal numbers, as follows.

Let $\alpha = \sharp(A)$ and $\beta = \sharp(B)$ be the cardinals of the sets A and B. In addition, let A be equivalent to a subset of B, that is, there is a function

$f : A \mapsto B$ which is one-to-one. Then write $A \preceq B$, which reads "A precedes B" and $\alpha \leq \beta$ (Lipschutz, 1968).

The notation $A \prec B$ is also used, to indicate that $A \preceq B$ and $A \nsim B$, that is, A is not equivalent to B. This implies, of course, that $\alpha < \beta$ means $\alpha \leq \beta$ and $\alpha \neq \beta$.

> **Cantor's Theorem**: for any set A, one has $A \prec 2^A$ and, therefore, $\alpha < \beta$. In which $\alpha = \sharp(A)$, so 2^α is the cardinal number of the family of subsets of A.

To prove Cantor's Theorem, it should be noted that the function $g : A \mapsto 2^A$, which sends each element $a \in A$ to the set consisting only of the element a, defined by $g(a) = \{a\}$, is bi-univocal, so $A \preceq 2^A$. It suffices, then, to show that A is not equivalent to 2^A.

Suppose the opposite is true, that there is a function $f : A \mapsto 2^A$ which is biunivocal and on. Let $a \in A$ be an atypical element, which is not a member of the set which is your image, that is, $a \notin f(a)$. Let B be the set of atypical elements, given by
$$B = \{x | x \in A, x \notin f(x)\}.$$

Note that B is a subset of A, that is, $B \in 2^A$. Therefore, because $f : A \mapsto 2^A$ is defined over 2^A, there is an element $b \in A$ with the property that $f(b) = B$. Thus, can b be considered typical or atypical? If $b \in B$, by the definition of B, $b \notin f(b) = B$, which is not possible. However, if $b \notin B$, then $b \in f(b) = B$, which is also impossible. Therefore, the initial assumption $A \sim 2^A$ led to a contradiction, from which it is concluded that it is false and the theorem is true (Lipschutz, 1968) (Braumann, 1987).

1.9 Georg Cantor

A paper written by Cantor, whose picture is shown in Figure 1.6, entitled "*Über eine Eigenschaft des Inbergriffes aller reellen algraishen Zahlen* (On a Property of the Totality of All Real Algebraic Numbers), published in a journal founded by the German mathematician August Leopold Crelle (1780-1855), who worked for the Prussian Ministry of the Interior on the construction and planning of roads and the one of the first railways in Germany, was considered a landmark in the history of Mathematics, and showed his disposition to place perceptive questions and obtain unexpected unconventional answers (Cantor, 1874) (Dunham, 2005).

Figure 1.6 A portrait of Georg Ferdinand Ludwig Philipp Cantor. Adapted from: Public Domain, https:// commons.wikimedia.org/w/index.php?curid=74820875

The *Journal für die Reine und Angewandte Mathematik* (Journal for Pure and Applied Mathematics), was commonly known as Crelle's Journal. It is known that Crelle befriended Niels Henrik Abel and published seven of Abel's papers in the first volume of his journal.

Cantor defined the cardinal and ordinal numbers and their arithmetic, but his theory of transfinite numbers was regarded as counter-intuitive by some mathematicians of this time, including his former professor Leopold Kronecker (1823–1891), a German mathematician who worked on number theory, algebra and logic. Kronecker publicly opposed Cantor, and attacked him personally, describing him as a "scientific charlatan", a "renegade" and a "corrupter of youth."

Cantor was also a musician and outstanding violinist, but had attacks of depression, starting from 1884 and recurring to the end of his life, which some have seen as a result of the hostile attitude of many of his contemporary mathematicians. But, those episodes could have also be caused by a bipolar disorder.

He participated in the founding of the German Mathematical Society, and chaired its first meeting in Halle, in 1891. At that first meeting, he introduced his diagonal argument and was elected as the first president of this society.

Georg Cantor retired from the University of Halle, in 1913, and lived in poverty and suffering from malnourishment during World War I. In 1917, he entered a sanatorium for the last time. There, the founder of the abstract set theory suffered a fatal heart attack, in 1918.

The Continuum Hypothesis, introduced by Cantor, was presented by David Hilbert (1862–1943) as the first of his twenty-three open problems in his address at the International Congress of Mathematicians, held in Paris, in 1900. Hilbert was a German influential mathematician who developed fundamental ideas in many areas, including invariant theory, the calculus of variations, commutative algebra, algebraic number theory, abstract geometry, spectral theory of operators and mathematical physics.

1.10 Problems

1. Discuss the notion that the universal set does not really contains all other sets, which is a known paradox attributed to Bertrand Arthur William Russel.

2. Prove that the complement of the limit superior of a sequence of sets is the limit inferior of the complement of that sequence, as follows

$$\overline{(\limsup A_n)} = \liminf \overline{A_n}.$$

3. Considering the following properties of a σ-algebra, prove that the empty set also belongs to the algebra.

   (a) $A \in \mathcal{B} \Rightarrow \overline{A} \in \mathcal{B}$
   (b) $A_i \in \mathcal{B} \Rightarrow \bigcup_{i=1}^{\infty} A_i \in \mathcal{B}$

   The properties guarantee the closure of the σ-algebra

4. Consider that the Cantor hotel receives an infinite number of trains, each one with an infinite number of tourists. Is it possible to accommodate them in the hotel apartments?

5. Prove that the cardinality of the power set $\mathcal{F} = 2^A$, that is, demonstrate that the family formed by all subsets of a set A, equals 2^α.

2
Relations and Functions

"A theory with mathematical beauty is more likely to be correct than an ugly one that fits some experimental data."
Paul Dirac

2.1 Definition of a Relation

A relation, sometimes called a binary relation, is a subset of the Cartesian product of sets that establishes a connection between two objects. In essence a mathematical relation is not very different from a human relation, but the rules are certainly better defined and more strict.

Given sets A and B, the Cartesian product $A \times B$, named in honor of René Descartes (1596–1650), a French philosopher, scientist and mathematician, is defined as $\{(a, b) \mid a \in A \text{ and } y \in B\}$, and its elements are called ordered pairs. That is, it represents a set of connected points.

A relation between the two sets is also defined as the collection of the ordered pair, in which the ordered pair is formed by the object from each set. A binary relation R over sets A and B is a subset of $A \times B$. The set A is called the domain, or set of departure of R, and the set B the range, or set of destination of R.

Because a relation is used to describe a connection between the elements of two sets, it maps the elements of the domain, to elements of the range, such that the resulting ordered pairs are of the form of input and output. When $A = B$, a relation is called homogeneous, or endorelation. Otherwise it is a heterogeneous relation.

Relations are used in many branches of Philosophy, Logic, Mathematics and Computer Science to model a wide variety of concepts, and include, among others:

- The "logic AND and OR relations" in Boolean logic;
- The "inclusion relation" in abstract set theory;

- The "congruence relation" in abstract algebra;
- The "greater than", "equal to", and "smaller than" relations in arithmetic;
- The "congruent to" relation in geometry;
- The "adjacent to" or "directed to" relations in graph theory;
- The "orthogonal to" and "parallel to" relations in linear algebra;

Furthermore, special types of relations that can be used to establish a correspondence between two quantities are known as functions. Therefore, a function is a subset, or a special kind, of a binary relation.

Example: It is possible to define the relation $R = \{(a, b) \in \mathbb{R} \times \mathbb{R} \mid a \geq b\}$. Thus, $(a, b) \in \mathbb{R} \times \mathbb{R}$ if and only if $a \geq b$. In this case, $(1, 2) \notin R$. Some authors use the notation aRb, to imply that a is related to b in some sense. ∎

2.1.1 Relation Representation

There are different ways to to express the relation, aside from set notation, that includes tables, plotting it on coordinate axes, or by a mapping, or Venn diagram. The type of representation depends on the features that are intended to be emphasized.

Example: The set $R = \{(1, 2), (2, 3), (3, 4), (4, 5), (5, 6)\}$ is a simple example of a relation between the sets $A = \{1, 2, 3, 4, 5\}$ and $B = \{2, 3, 4, 5, 6\}$, which is usually written in set notation form with curly brackets. This representation is also called roster form. The relation can also be expressed, more compactly, in set-builder form, as $R = \{(a, b) : b = a + 1, a \in A, b \in B\}$. Note that $3 < 4$, but $4 \not< 3$, which demonstrates that a is related to b does not necessarily imply that b is also related to a. ∎

The domain of a relation $R \subseteq A \times B$ is defined as

$$\operatorname{dom} R = \{a \in A \mid (a, b) \in R \text{ for some } b \in B\}, \tag{2.1}$$

and the image, or range, is specified as

$$\operatorname{im} R = \{b \in B \mid (a, b) \in R \text{ for some } a \in A\}. \tag{2.2}$$

The mentioned relation is represented, apart from the set notation, by using a tabular notation, as in Table 2.1, as a plotting on coordinate axes, as in Figure 2.1, or as a mapping, or arrow, diagram, as depicted in Figure 2.2.

Because relations are sets, they can be manipulated using set operations, including union, intersection, and complementing, and must satisfy the laws of an algebra of sets. In addition, operations such as the converse of a relation

2.1 Definition of a Relation

Table 2.1 Relation in tabular form.

a	b
1	2
2	3
3	4
4	5
5	6

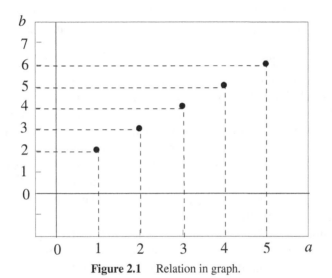

Figure 2.1 Relation in graph.

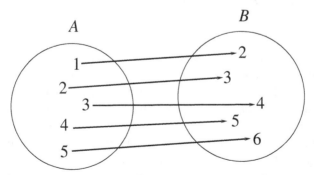

Figure 2.2 Relation expressed as a mapping, or Venn diagram.

and the composition of relations are possible, and satisfy the laws associated to the algebraic logic.

2.1.2 Types of Relations

Relations are generalizations, or supersets, of functions. A relation states that the elements from two sets A and B are connected in a certain way. More formally, a relation is defined as a subset of $A \times B$. Different types of relations are described as follows:

- Empty Relation – When there is no element of set A that is related, or mapped, to any element of B, then the relation R in A is an empty relation, and also called the void relation, that is $R = $. For example, $A = \{1, 2, 3\}$ and the relation on A, $R = \{(a, b)$, in which $a + b = 10\}$. This is an empty relation, as no two elements of A are added up to 10.
- Universal relation – If all the elements that belong to one set are mapped to all the elements of another set, or to itself, then such a relation is known as a universal relation. Consider that R is a relation in a set A. Then it is a universal relation because, in this full relation, every element of A is related to every element of A. that is $R = A \times A$. For example, $A = \{1, 2, 3\}$, $B = \{4, 5, 6\}$ and $R = \{(a, b)$ in which $x < y\}$.
- Identity Relation – If every element of set A is related to itself only, it is called an identity relation. It is written as $I = \{(a, a) : \text{ for all } a \in A\}$. For example, $P = \{1, 2, 3\}$ then $I = \{(1, 1), (2, 2), (3, 3)\}$.
- Inverse Relation – If R is a relation from set A to set B, that is $R = A \times B$. On the other hand, the inverse relation $R^{-1} = \{(b, a) : (a, b) \in R\}$.
- Reflexive Relation – A relation is a reflexive relation if every element of set A maps to itself, that is, for every $a \in A$, $(a, a) \in R$. For example, $A = \{1, 2\}$ then $R = \{(1, 1), (2, 2)\}$ is a reflexive relation.
- Symmetric Relation – A symmetric relation is a relation R on a set A if $(a, b) \in R$ then $(b, a) \in R$, for all $a, b \in A$. Consider that $P = \{1, 2\}$, then a symmetric relation is $R = \{(1, 2), (2, 1)\}$.
- Transitive Relation – If $(a, b) \in R$, $(b, c) \in R$, then $(a, c) \in R$, for all $a, b, c \in A$ then this relation in the set A is transitive. For example, suppose there is a set $A = \{a, b, c\}$, then a transitive relation is $R = \{(a, b), (b, c), (c, a)\}$.
- Equivalence Relation – If a relation is reflexive, symmetric and transitive, then the relation is called an equivalence relation.
- One to One Relation – In a one to one relation each element of one set is mapped to a distinct element in another set. For example, suppose there

are two sets $A = \{1, 2, 3, 4\}$ and $B = \{a, b, c, d\}$. Then a one to one relation can be $R = \{(1, a), (2, b), (3, c), (4, d)\}$.
- One to Many Relation – In a one to many relation, a single element of one set is mapped to more than one element in another set. For example, given two sets $A = \{1, 2, 3, 4\}$ and $B = \{a, b, c, d\}$, a one to many relation is written as $R = \{(1, a), (1, b), (1, c), (1, d)\}$.
- Many to One Relation – If more than one element of one set are mapped to a single distinct element of another set then such a relation is referred to as many to one relation. For example, $A = \{1, 2, 3, 4\}$ and $B = \{a, b, c, d\}$, then $R = \{(1, a), (2, a), (3, a), (4, a)\}$ is a many to one relation.
- Many to Many Relation – In a many to many relation, one or more elements of one set is mapped to the same or a different element of another set. If $A = \{1, 2, 3, 4\}$ and $B = \{a, b, c, d\}$, then one has that $R = \{(2, a), (3, a), (2, c), (4, c)\}$ is an example of a many to many relation.

2.2 Definition of Function

A function is a relation for which there is only one output for each input. It is a special kind of relation, a set of ordered pairs, which follows a certain prescribed rule. A function is also defined as an expression, rule, or law that establishes a relationship between one variable, called the independent variable, and another variable, that is the dependent variable.

Functions are used in all branches of Mathematics and are essential to formulate physical relationships in several other sciences. The relationship between the input and output variables is symbolized as $y = f(x)$, and it indicates that for every x, there is a unique value of y (Britannica, 2021). Figure 2.3 illustrates the relationship between the set of relations and the set of functions.

A function is characterized by a domain, A, and a range, B, along with a formula or description of the relationship between the elements of these sets.
- Domain – Collection of the first values in the ordered pair, that is, the set of all input $x \in A$ values,
- Range – Collection of the second values in the ordered pair, that is, the set of all output $y \in B$ values.

2.2.1 Types of Functions

The mathematical functions are determined based on the domain, range and defining expression. In terms of relations, it is possible to characterize the functions as (Wikipedia, 2022):

- A one to one function, or injective function – A function is injective if $f : A \to B$ is said to be one to one if for each element of A there is a distinct element of B.
- A many to one function – A function which maps two or more elements of A to the same element of set B.
- An onto function, or surjective function – A function for which every element of set B there is pre-image in set A
- A one-one correspondence or bijective function – The function f matches with each element of A with a discrete element of B and every element of B has a pre-image in A.

The function f, whose domain is the set X, is said to be injective if for all x and y in X, if $f(x) = f(y)$, then $x = y$. Therefore, $f(x) = f(y)$

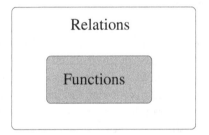

Figure 2.3   A function is a subset of a relation.

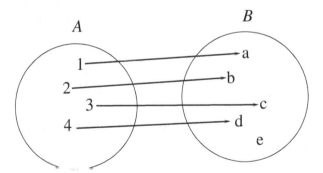

Figure 2.4   Venn diagram for an injective function.

2.2 Definition of Function

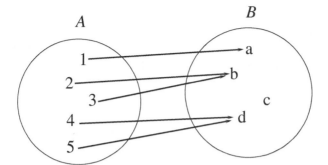

Figure 2.5 Venn diagram for a many to one function.

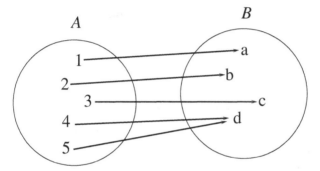

Figure 2.6 Venn diagram for a surjective function.

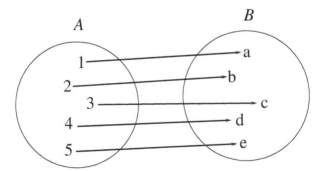

Figure 2.7 Venn diagram for a bijective function.

implies $x = y$. Some properties of the injective functions are presented in the following.

- If the functions f and g are both injective then $f \circ g$ is injective.

- If the composite function $g \circ f$ is injective, then f is injective, but g is not necessarily injective.
- The function $f : X \to Y$ is injective if and only if, given any functions $g, h : Z \to X$, whenever $f \circ g = f \circ h$, then $g = h$.
- If the function $f : X \to Y$ is injective and A is a subset of X, then the inverse function $f^{-1}(f(A)) = A$. Therefore, the set A can be obtained from its image $f(A)$.
- If the function $f : X \to Y$ is injective and A and B are both subsets of X, then $f(A \cap B) = f(A) \cap f(B)$.
- Every function $h : W \to Y$ can be decomposed as $h = f \circ g$ for a suitable injection f and a surjection g. This decomposition is unique.
- If $f : X \to Y$ is an injective function, then Y has at least as many elements as X, considering the cardinal numbers associated to the sets.
- If the sets X and Y are finite with the same number of elements, or same cardinality, then $f : X \to Y$ is injective if and only if f is surjective. In this case f is also bijective.

Example: The following functions are injective:

- The affine function $f : \mathbb{R} \to \mathbb{R}$, defined by $f(x) = ax + b$ is injective.
- The quadratic function $g : \mathbb{R} \to \mathbb{R}$, defined by $g(x) = x^2$ is not injective, because $g(a) = a^2 = g(-a)$. However, g if is defined in a way that its domain is the non-negative real numbers $[0, +\infty)$, then g becomes injective.
- The exponential function $\exp : \mathbb{R} \to \mathbb{R}$, defined by $\exp(x) = e^x$, is injective, but not surjective, because no real number maps to a negative value. ■

A function is surjective if each element of the codomain is mapped to by at least one element of the domain. That is, the image and the codomain of the function are equal. In other words, $\forall y \in Y, \exists x \in X$ such that $y = f(x)$. Some properties of the surjective functions are presented in the following.

- A function $f : X \to Y$ is surjective if and only if it is right-invertible, that is, if and only if there is a function $g : Y \to X$ such that $f \circ g$ is the identity function on Y.
- The composition of two surjections is also a surjection. But, if $g \circ f$ is surjective, it is only possible to conclude that g is surjective

As previously implied, a function is bijective if it is both injective and surjective. A bijective function is called a bijection. A function is bijective if and only if every possible image is mapped to by exactly one domain element. This similar condition is mathematically expressed as, the function $f\colon X \to Y$ is bijective if for all $y \in Y$, there is a unique $x \in X$ such that $f(x) = y$. Some properties of the bijective functions are presented in the following.

- A function $f\colon X \to Y$ is bijective if and only if it is invertible, that is, there is a function $g\colon Y \to X$ such that $g \circ f$ is the identity function on X and $f \circ g =$ is the identity function on Y. This surjective function maps each image to its unique preimage.
- The composition of two bijections is a bijection, but if $g \circ f$ is a bijection, then it can only be concluded that f is injective and g is surjective.

Example: It is possible to give a simple classification to understand the types of functions that are commonly found in most mathematical problems:

- Algebraic functions
    - Polynomial functions
        * Constant, linear, quadratic and cubic functions
    - Rational, radical and piecewise functions
- Transcendental functions
    - Exponential functions
    - Logarithmic functions
    - Trigonometric functions
    - Hyperbolic functions ■

2.3 Mathematical Functions

A function is a rule that associates an element of a set $a \in A$, called a domain, to another element of a set $b \in B$, known as the range. When the operation is made with real numbers, x and y, the function is said to be real, and $y = f(x)$ is written. In In general, f defines a mapping from A to B, as illustrated in Figure 2.8.

If a is an element of the set A, the corresponding element $b = f(a)$ is called the image of a by mapping f, therefore a is called the pre-image of b. Generally, b can have multiple pre-images, and B can contain elements that do not have pre-image. If b has only one pre-image, this pre-image is denoted by $f^{-1}(b)$.

If C is a subset of A, the set of all elements $f(a) \in B$ such that $a \in A$ is called the image of C, denoted by $f(C)$, as shown in Figure 2.9. The set

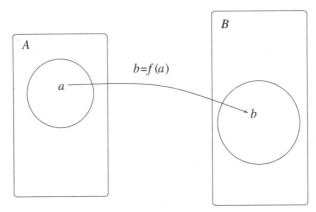

Figure 2.8 A function associates the element $a \in A$, in the domain, with the element $b \in B$, in the co-domain.

of all elements of A whose images belong to a given set $D \subset B$ is called a pre-image of D, and written as $f^{-1}(D)$. If no element has a pre-image, then $f^{-1}(D) = \emptyset$.

A function is said to be one-to-one, or maps A to B, if $f(A) \subset B$, as usual. In case it maps A over B, that is, if $f(A) = B$, it is called a surjection. A function is a bijection, or one-to-one, if each element $b \in B$ has a unique pre-image $f^{-1}(b)$, and f is said to establish a one-to-one correspondence

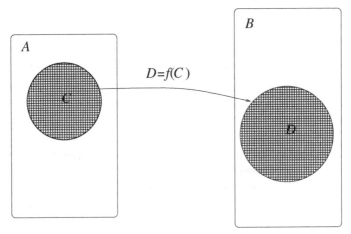

Figure 2.9 A function maps the set $C \subset A$ to its image $D \subset B$.

between A and B, and the mapping f^{-1}, for each $b \in B$, is called the inverse function of f, as illustrated in Figure 2.10.

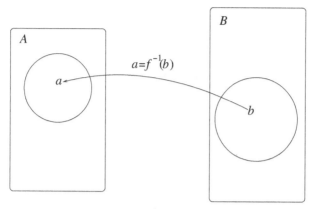

Figure 2.10 The inverse function associates the element $b \in B$, of the range, with the element $a \in A$.

2.3.1 Indicator Function

A very convenient operation with sets, that is useful for characterizing the probability measure, is the indicator function $I_A(x)$, defined for a set A of a space Ω. It is also known as the characteristic function (Magalhães, 2006):

$$I_A(x) = \begin{cases} 1, & \text{for } x \in A \\ 0, & \text{for } x \notin A. \end{cases} \quad (2.3)$$

The indicator function has the following properties (Marques, 2009):

1. A set is completely determined by its indicator function,

$$A = \{\omega : I_A(\omega) = 1\}. \quad (2.4)$$

2. For the sets $A, B \subset \Omega$,

$$I_A \leq I_B \Leftrightarrow A \subset B. \quad (2.5)$$

3. For the equal sets A and B,

$$I_A = I_B \Leftrightarrow A = B. \quad (2.6)$$

4. The indicator of the empty set and of the universal set,

$$I_\emptyset = 0 \text{ and } I_\Omega = 1. \tag{2.7}$$

5. The set intersection indicator. If $B = \cap_{k=1}^n A_k$,

$$I_B = \prod_{k=1}^n I_{A_k}. \tag{2.8}$$

6. The indicator of the union of disjoint sets. If $B = \cup_{k=1}^n A_k$, $A_i \cap A_j = \emptyset$, for $i \neq j$,

$$I_B = \sum_{k=1}^n I_{A_k}. \tag{2.9}$$

7. The indicator of the symmetric difference of sets. If $C = A \Delta B$,

$$I_C = |I_A - I_B|. \tag{2.10}$$

8. Considering the module 2 operation, the indicator of the symmetric difference of sets A and B provides, for $C = A \Delta B$,

$$I_C = I_A + I_B \pmod{2}. \tag{2.11}$$

9. Given a set A, its complement with respect to the universal set Ω has the indicator function,

$$I_{\overline{A}}(x) = 1 - I_A(x). \tag{2.12}$$

This function is helpful in the operational definition of sums and integrals used in probability measure, as it delimits the set in which these measures operate. It is also used to perform a probability approach by means of the expected value operator, to be defined.

2.3.2 Fuzzy Sets

The indicator function is the basis for the construction of fuzzy sets by Lotfi Aliasker Zadeh (1921–2017), a mathematician, electrical engineer, computer scientist, artificial intelligence researcher, and professor of computer science at the University of California, Berkeley. He was born in the Azerbaijan (Zadeh, 1965).

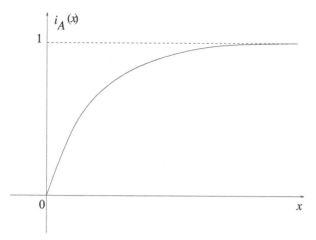

Figure 2.11 Example of a fuzzy indicator function.

Given a space of points (objects) X, a class of fuzzy sets A in X is characterized by the membership relationship defined by the generic function indicator $i_A(x)$, which associates to each point in X a real number in the range $[0, 1]$, with the value of $i_A(x)$ representing the degree of pertinence of the element x in A.

Figure 2.11 illustrates the concept of membership for the case of fuzzy sets. The closer the value of $i_A(x)$ is of unity, the higher the degree of membership of x in A. When A is a set in the ordinary sense, its indicator function can only assume the values 1 and 0, indicating whether or not the element x belongs to the set A. The ordinary, or conventional, sets may be called fussy sets.

Example: An interesting application of the indicator function is obtained when considering the set of real numbers $\mathbb{R}$ and the subset A of the non-negative numbers. The indicator function takes the form

$$I_A(x) = \begin{cases} 1, & \text{if } x \geq 0; \\ 0, & \text{if } x < 0. \end{cases} \blacksquare$$

This function is the well-known unit step, usually denoted by $u(x)$, and represents various engineering operations, including the switching on and off of current and voltage sources. As a mathematical function, the unit step is used to mark the domain of other functions. For example, the composition $u(x + a) - u(x - a)$ delimits the range $[-a, a]$.

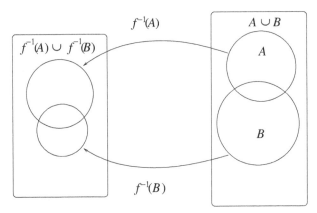

Figure 2.12 The inverse function of the union of sets A and B is the union of the inverse functions of A and B.

2.3.3 Properties of Set Functions

Some properties of set functions are useful in deducing the results of operation between sets. The pre-image of the union of two sets is the union of the pre-images of the sets, that is,

$$f^{-1}(A \cup B) = f^{-1}(A) \cup f^{-1}(B). \qquad (2.13)$$

If $x \in f^{-1}(A \cup B)$, then $f(x) \in A \cup B$, therefore, $f(x)$ belongs to, at least, one of the sets A or B. But, then x belongs to, at least, one of the sets $f^{-1}(A)$ or $f^{-1}(B)$, that is, $x \in f^{-1}(A) \cup f^{-1}(B)$, as shown in Figure 2.12.

On the other hand, if $x \in f^{-1}(A) \cup f^{-1}(B)$, then x belongs to at least one of the sets $f^{-1}(A)$ or $f^{-1}(B)$. Therefore, $f(x)$ belongs to at least to one of the sets A or B, that is, $f(x) \in A \cup B$, that is, $x \in f^{-1}(A \cup B)$.

In the same way, it is shown that the pre-image of the intersection of two sets is the intersection of pre-images of sets,

$$f^{-1}(A \cap B) = f^{-1}(A) \cap f^{-1}(B). \qquad (2.14)$$

If $x \in f^{-1}(A \cap B)$, then $f(x) \in A \cap B$, so $f(x)$ belongs to both sets A and B. But then x belongs to both sets $f^{-1}(A)$ and $f^{-1}(B)$, that is, $x \in f^{-1}(A) \cap f^{-1}(B)$.

However, if $x \in f^{-1}(A) \cap f^{-1}(B)$, then x belongs to both sets $f^{-1}(A)$ and $f^{-1}(B)$. So $f(x)$ belongs to A and simultaneously to B, that is, $f(x) \in A \cap B$, which means, $x \in f^{-1}(A \cap B)$.

2.3 Mathematical Functions

The image of the union of two sets is the union of the images of the sets, as illustrated in Figure 2.13,

$$f(A \cup B) = f(A) \cup f(B). \tag{2.15}$$

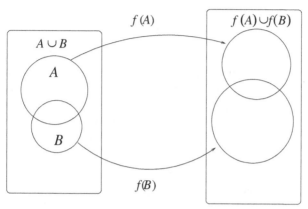

Figure 2.13 The union function of the sets A and B is the union of the functions of A and B.

If $y \in f(A \cup B)$, then $y = f(x)$, in which x belongs to at least one of the sets A or B. Therefore, $y = f(x)$ belongs to at least one of the sets $f(A)$ or $f(B)$, that is, $y \in f(A) \cup f(B)$.

The converse is also true, since if $y \in f(A) \cup f(B)$, then $y = f(x)$, in that x belongs to at least one of the sets A or B. Therefore, $x \in A \cup B$ and, consequently, $y = f(x) \in A \cup B$.

Finally, the image of the intersection of two sets is not necessarily the intersection of the two images from the sets (Kolmogorov and Formin, 1970). For example, consider the function $f : \mathbb{R}^2 \mapsto \mathbb{R}$, which maps the xy coordinate plane onto the abscissa axis, projecting the point (x, y) at the point $(x, 0)$. In this case, the line segments $0 \leq x \leq c$, $y = a$ and $0 \leq x \leq c$, $y = b$ do not intercept, although their images match.

Example: If A is the set of values of a function, the *supremum* of A, denoted sup A can be specified in several ways.

Suppose A is the set of all $f(x)$ values corresponding to $x > 0$, in which

$$f(x) = \frac{x-1}{x},$$

so sup $A = 1$, which can be written as

$$\sup A = \sup_{x>0} f(x) = 1.$$

Since $A = \{f(x) : x > 0\}$, one can still write as

$$\sup\left\{\frac{x-1}{x} : x > 0\right\} = 1. \blacksquare$$

2.4 The Count of Arts and Mathematics

Johan Maurits van Nassau-Siegen (1604-1679), whose picture is shown in Figure 2.14 also know by his Portuguese name João Maurício de Nassau Siegen, was born in Dillenburg, Germany, and received a good education in Basel, where he reached the age of 10. In 1616, he entered the Collegium Mauritianum. He was called "the Brazilian" for his fruitful period as governor of Dutch Brazil, he was Count and (from 1664) Prince of Nassau-Siegen.

In 1621, as was usual among members of the nobility, he entered a military career in the service of the Netherlands, at the time of the Thirty

Figure 2.14 A portrait of Johan Maurits van Nassau-Siegen. Adapted from: Public Domain, https://commons.wikimedia.org/w/index.php?curid=2145765

Years' War against Spain. He was a knight of the Order of Saint John and carried out the military campaign of (Breda, 1625), to retake the city from the Spanish. From 1626 he had been promoted to captain and, in 1629, to colonel.

In 1632, Maastricht was taken, and Nassau distinguished itself in the siege that culminated in the conquest of Nieuw Schenckenschans, on an island in the River Rhine, in 1636, confirming his prestige and military experience. The victory had repercussions and made his name well respected.

From this period onwards Nassau developed contact with artists, mathematicians and philosophers. Among them, the great French mathematician, physicist and philosopher René Descartes (1596-1650), and the engineer Simon Stevin (1548-1620), one of the great mathematicians of the 17th century, a clerk in Bruges, Belgium, who was his tutor.

The Cartesian product, popularized on account of Descartes' Latin name Renatus Cartesius, was originally the brainchild of philosopher, physicist, astronomer, biologist and mathematician Nicole Oresme (1323-1382), Bishop of Lisieux. This product is a way of expressing the idea of set indexing, which expands the possibilities of using set theory and allows the production of entities known as vectors and signs.

An ordered pair is unambiguously defined using set notation such as $(x, y) = \{x, \{x, y\}\}$, in which x is the first element of the pair, typically the independent variable, and the second element y represents the dependent variable.

Descartes, who graduated in law, in 1616, from the University of Poitiers, was noted for his revolutionary work in philosophy and science, and for suggesting the fusion of algebra with geometry, which created analytic geometry. He is also seen as the founder of modern philosophy, and considered one of the most important thinkers in the history of Western thought, for introducing rationalism into philosophy.

Despite having a law degree, Descartes never practiced law and, in 1618, he went to Holland, where he enlisted in Nassau's army, with the intention of pursuing a military career. He served for a time with Nassau but, encouraged by a friend, moved to Germany to continue his military service there.

Eventually Nassau was hired by the West India Company and sent to Brazil, in 1637, to govern the province of Pernambuco, which had been conquered by the Dutch. The Dutch destroyed the beautiful city of Olinda, and made Recife the capital of their tropical empire. Nassau hired the famous architect Pieter Post of Haarlem to transform Recife, by building a new town adorned with public buildings, bridges, channels, and gardens in the then Dutch style, later naming the newly reformed town Mauritsstad.

2.5 Problems

1. Define the relation $R = \{(a, b) \in \mathbb{R} \times \mathbb{R} \mid a - b < 0\}$, that is, $(a, b) \in \mathbb{R} \times \mathbb{R}$ if and only if $a - b < 0$. Give a case in which $(a, b) \notin R$.

2. Verify if the cubic function $g : \mathbb{R} \to \mathbb{R}$, defined by $g(x) = x^3$ is injective.

3. Prove that the set intersection indicator for $B = \cap_{k=1}^{n} A_k$ is given by

$$I_B = \prod_{k=1}^{n} I_{A_k}.$$

4. Considering the set of real numbers $\mathbb{R}$, compute the indicator function of the sine function $\sin x$.

5. Consider that A is the set of all $f(x)$ values corresponding to $x > 0$, in which

$$f(x) = \frac{x-1}{x},$$

and determine the *infimum* of A, that is, $\inf A$.

3

Fundamentals of Measure Theory

"Measure what is measurable, and make measurable what is not so."
Galileo Galilei

3.1 Measuring History

Counting is one of the main activities of the human species, who have done it since ancient times. It started with finger counting, one of the reasons the base 10 system is universal, and proceeded to abstract measure. But it is not an exclusive feature of humans. Most animals have the ability to count, but only humans display the capacity for abstraction, the ability to represent measurement in symbolic form.

The Egyptians developed tools to measure areas, which was important to estimate quantity of crops produced, obtained thanks to the Nile floods. But the first civilization to systematically study geometrical figures and compute areas were the Greeks. Euclid of Alexandria (c.360–c.295 BC), Eudoxus of Cnidus (390–338 BC), Archimedes of Syracuse (287–c.212 BC) and other mathematicians proved fundamental theorems that led to the computation of areas, and ultimately to the development of Calculus.

Eudoxus made a remarkable contribution to Mathematics by his seminal work on integration, using his method of exhaustion, which allowed approximation of two unequal quantities by the reduction of their difference. This method was developed as a direct result of his work on the theory of proportion. He also considered numbers as the ratio of lengths, which led to the discovery of rational numbers and, later, to the comparison of irrational numbers.

Archimedes of Alexandria was born in Syracuse. He was a mathematician, a physicist, an engineer, an inventor, an astronomer, and one of the leading scientists of classical antiquity. He came very close to discovering Calculus, by applying concepts of infinitesimals and the method of exhaustion, and he did it nineteen centuries before Isaac Newton and Gottfried Leibniz were born. Archimedes believed that everything could be measured. He improved the Greek number system and computed the area of the circle, using Eudoxus' method of exhaustion (Alencar, 2008a).

Zeno of Elea (495–430 BC), a member of the Eleatic School, founded by Parmenides (c.515–c.450 BC), and the inventor of the dialectic, according to Aristotle (384–322 BC), also created four paradoxes involving the concept of infinity, which could not be explained by the logic of his time. The explanation required modern concepts of convergence of series and limits.

Four centuries ago, Johannes Kepler (1571–1630) discussed a method for computing the areas of plane figures and volumes of solids using infinitesimal elements, probably based on the work of Eudoxus and Archimedes.

Galileo Galilei (1564–1642) developed an integration method to prove that, for constant acceleration, the area under the velocity curve was the distance covered. Bonaventura Cavalieri (1598–1647) also used the ideas of Kepler and Galileo to study plane figures, using indivisibles.

The independent work of Isaac Newton (1642–1727), in Great Britain, and Gottfried Wilhelm Leibniz (1646–1716), in Germany, led to the development of Calculus. They concluded that integration could be thought of as the inverse of differentiation.

Leibniz was studying sequences of functions, when he discovered the way to integrate and differentiate. He also created the elongated "S", as the integral symbol, and used "d" as the symbol for the differential. Newton imagined the fluxions to account for the indivisibles, or increments, and used the dot as a notation for differentiation.

Augustin Louis Cauchy (1789–1857) gave a rigorous treatment to the integral, and established the integral of a continuous function as the sum of the product of values of the function by the partitions of the independent variable in the measured interval (Alencar, 2008b).

Georg Friedrich Bernhard Riemann (1826–1866), a German mathematician with fundamental contributions to analysis and differential geometry, generalized Cauchy's integration method and studied trigonometric series. His method of integration is widely used in Engineering, Physics and Mathematics nowadays.

Jean-Gaston Darboux (1842–1917), a French mathematician who worked on orthogonal surfaces, demonstrated that a function can be integrated, or has a measurable area, when the superior and inferior Riemann sums converge to the same value, as the intervals go to zero, for any partition.

Johann Peter Gustav Lejeune Dirichlet (1805–1859), a German mathematician, who made important contributions to number theory, including the field of analytic number theory and the theory of Fourier series, aside from other topics in mathematical analysis, discovered certain functions that could not be integrated in the Riemann sense, or using the Riemann measure.

Dirichlet presented a function that assumed unit value when the independent variable belonged to the set of rational numbers, and zero value for points in the set of irrationals. The union of the irrational and rational sets produces the real line. Dirichlet's function has an infinite number of discontinuities, and can not be measured using Riemann's integral.

The explanation is the following. The set of rationals has the same cardinality, or can be put in a one to one correspondence with the set of integers, because it is countable. On the other hand, the set of irrationals is equivalent, or has the same cardinality, to the set of real numbers, which is continuous, or uncountable.

Marie Ennemond Camille Jordan (1838–1922), a French mathematician, known for his fundamental work in group theory and analysis, and Giuseppe Peano (1858–1932), an Italian mathematician and glottologist, founder of mathematical logic and set theory, proposed the useful idea of internal and external content of a set, which was fundamental to the development of a formal notion of measure.

Félix Edouard Justin Émile Borel (1871–1956) devised the first measure theory for a set of points. He defined the content of a set as its measure, based on the research of Georg Cantor (1845–1918). He also defined the measure of a union of sets as the sum of the individual measures, and proved that sets with nonzero measure are uncountable, that is, they cannot be put in a one to one, or bijective, correspondence with the set of integers.

It is interesting to note that the measure of an interval that contains only one point is zero, as the point has no dimension. A finite set of points has also zero measure, because it is a sum of null measures. The idea can be generalized to a denumerable (countable) set, even if it is infinite, because the infinite, but enumerable, sum of zero measures is also zero.

Cantor studied infinite, uncountable and continuous sets. He also devised even larger infinites, and created the transfinite numbers to manage them. But the measurement of those sets, and of the functions defined on them, was done by Henri Léon Lebesgue (1875–1941), a French mathematician (Alencar, 2008c).

Henri Lebesgue was born in the city of Beauvais, France. He devised an original integration method, different from Riemann's, published in the French journal *Comptes Rendus*, in April, 1901. In the following year, he defended his doctoral thesis in the Faculté des Sciences de Paris. He was influenced by the work of René-Louis Baire (1874–1932) on discontinuous functions.

Lebesgue's method to integrate a function is to measure all infinitesimal intervals (indivisibles) that correspond to a certain value of the function. The individual measures are then added. This is analogous to using the function differential, instead of the independent variable differential.

Set Theory, created by Cantor, together with Measure Theory, developed by Lebesgue, formed the basis for Probability Theory, an axiomatic construction by Andrei Kolmogorov (1903–1987) (Alencar, 2008d).

3.2 Measure in an Algebra of Sets

Weight, mass, distance and area have been important measures since ancient times, but their nature is different from the one of counting objects. The continuous nature of an area, for instance, requires another form of evaluation. In Egypt, because of the harvesting of the land along the Nile river, the Pharaohs had much interest in the measurement of the area invaded by the Nile's floods.

The ancient Egyptians developed the nilometer, a method to measure the Nile's flood level, as their harvests and livelihood depended on the river's annual flow. It was a method to record the level of a flood using marks on river banks, along stairs leading to the river, on stone pillars or in water wells. These measurements were used to estimate crop yields and taxes to be collected by the Pharaohs (Toussoun, 1925). Measurements like that attribute a real number to a set, in a given collection of sets.

In an algebra, or σ-algebra, a measure is a non-negative real function μ in $\mathcal{F}$, such that $A_1, A_2, A_3, \ldots$ form a finite or denumerable collection of disjoint sets which belong to $\mathcal{A}$, that is (Ash, 1990)

$$\mu\left(\bigcup_n A_n\right) = \sum_n \mu(A_n). \tag{3.1}$$

A measure is always applied to a set, which represents a certain feature of the elements. For example, there is no sense in measuring an object, unless some of its properties are pointed out, such as weight, mass, hardness, elasticity, conductivity, resistance, color or size. Therefore, it is not possible to measure an element, but only the set that contains the element.

The set of real numbers, $\mathbb{R}$, is used to measure distance, voltage, current, power, while the area, or statistics, requires the set $\mathbb{R}^2$, and a volume measurement or mass is done over $\mathbb{R}^3$. Of course, there are several other applications of measures, which include all fields of science. Measuring is one of the main activities of humankind.

The formal notion of measure first appeared in the theory of real functions, but it has been used in many branches of Mathematics, which include functional analysis, probability, information theory and dynamic systems theory (Kolmogorov and Formin, 1970). Common sense requires that a measure must be real and non-negative, as well as additive. For example, if an object is divided into two parts, the sum of the masses of each part equals the mass of the original object, and the mass is real and non-negative.

The measure of the empty set is zero, again as required by common sense. If the measure of the universal set is unit, that is, $\mu(\Omega) = 1$, then μ is a probability measure.

Suppose, for instance, that the objective is to measure the area of a sandlot $A \subset \Omega$, in which Ω is the set of all available areas. Then, at first, it is necessary to establish the total area of the sandlot $\mu(\{A\})$, or simply, $\mu\{A\} = \alpha$. One must notice that, if there is no area, the area is null, that is, $\mu\{\emptyset\} = 0$.

Besides, if the sandlot is divided into two parts, B and C, the total area must be the sum of the individual parts, $\mu(\{B\} \cup \{C\}) = \mu(\{B\}) + \mu(\{C\}) = S$, because the sets are disjoint, and do not occupy the same space.

Example: Consider Ω any universal set, and $\mathcal{A}$ the family of all subsets of Ω. Define $\lambda(A)$ as the number of points of A. Therefore, if A has N members, $N = 0, 1, 2, 3, \ldots$, then $\lambda(A) = N$. If A is an infinite set, $\lambda(A) = \infty$. The set function λ is a measure on $\mathcal{A}$, known as the counting measure on Ω. ∎

Several techniques were developed to measure, and a set of rules was established, known as the International System of Units (SI), to standardize the measurement process everywhere. The SI was established and is maintained by the General Conference on Weights and Measures (CGPM); it is the system of measurement with official status in most countries, employed in science, technology, industry, and commerce.

On the other hand, the integral is used in most exact sciences to compute complex areas, thus, it is necessary to define formal rules for the integral calculus, considering that not all sets can be measured using the conventional methods of computation. In fact, some sets can be defined in quite strange ways. Those sets are the preferred object of study of most mathematicians.

Example: For a rectangle of edges $A = [a, b] \times [c, d]$ the measure of the area is given by $\mu(A) = (c - a) \cdot (d - b)$. This formula can be generalized to any dimension. ∎

A function with domain in a class, or family, $\mathcal{A}$, of subsets from a space Ω, is called a set function. The measure μ is a set function with many properties. It can be monotonically non-decreasing if (Marques, 2009)

$$\mu(A) \leq \mu(B), \text{ for } A \subset B, A, B \in \mathcal{F}, \qquad (3.2)$$

or monotonically non-increasing if,

$$\mu(A) \geq \mu(B), \text{ for } A \subset B, A, B \in \mathcal{F}. \qquad (3.3)$$

For a measure μ defined on a σ-algebra $\mathcal{A}$ of subsets from an abstract nonempty space, Ω, the triple $(\Omega, \mathcal{A}, \mu)$ is called a measure space. If, in addition, $\mu(\Omega) = 1$, the triple $(\Omega, \mathcal{A}, \mu)$ is called a probability space, and μ is the probability measure of the space.

Example: If the measure $\mu(B)$ is finite and nonzero, it is possible to define a probability measure as,

$$\nu_B(A) = \frac{\mu_B(A)}{\mu(B)} = \frac{\mu(A \cap B)}{\mu(B)}, \forall A, B \in \mathcal{F}. \blacksquare \qquad (3.4)$$

Example: The contraction is a useful measure. For the measurable sets $A, B \in \mathcal{A}$, one can define the contraction of the measure of A into B, as

$$\mu_B(A) = \mu(A \cap B), \qquad (3.5)$$

which has the properties:

$$\mu_B(\overline{B}) = 0, \qquad (3.6)$$

and

$$\mu_B(A) = \mu(A), \forall A \in \mathcal{F}, A \subset B. \blacksquare \qquad (3.7)$$

3.3 The Riemann Integral

Bernhard Riemann (1826–1866) developed the integral that bears his name to establish with rigor the exhaustion method, created by Eudoxus of Cnido (390–338 BC) and later improved by Archimedes (287–212 BC). Isaac Newton (1642–1727) and Gottfried Wilhelm Leibniz (1646–1716) gave Calculus its present form.

Some mathematicians, such as Blaise Pascal (1623–1662), had dealt with integration before Newton. Leibniz, for instance, was influenced by Pascal's work with infinitesimals. The notation used by Leibniz for the derivative of a product of two factors is based on Pascal's way of expressing it, as $(x + \mathrm{d}x)(y + \mathrm{d}y) - xy = x\mathrm{d}y + y\mathrm{d}x$ (the term $\mathrm{d}x\mathrm{d}y$ is discarded as it is incomparably less than $x\mathrm{d}y + y\mathrm{d}x$). At that time, the integral was seen as the antiderivative, but Riemann removed this constraint (Struik, 1987).

The Riemann integral, which is familiar to the engineering students, applies to continuous functions that do not present too many points of discontinuity. Although the Riemann integration is a successful attempt to provide such a firm mathematical foundation for the subject, it does not deal very well with taking limits of sequences of functions, which makes such limiting processes difficult to analyze.

Therefore, it is not possible to use the Riemann integral for a generic measurable function, for example. In fact, a function f can be discontinuous everywhere, or, maybe, there is no sense in saying that a function is continuous if it is defined for an abstract set (Kolmogorov and Formin, 1970).

The first rigorous treatment of the integral was given by Augustin Louis Cauchy (1789–1857), and Jean-Baptiste Joseph Fourier (1768–1830) noticed that the definition in terms of the derivative was too restrictive.

Fourier published an article on the diffusion of heat, in 1807, in which he discussed the representation of sets of functions by trigonometric series, and managed to prove that the solution of the equation

$$f(x) = a_0 + \sum_{n=1}^{\infty} a_n \cos nx + b_n \sin nx, \tag{3.8}$$

for an infinite number of unknown coefficients, could be expressed in terms of integrals. He also proved that the solution exists even for discontinuous functions, whose anti-derivatives cannot be found.

Riemann submitted his *Habilitationsschrift* (probationary essay) for admission to the faculty in 1853, entitled "On the Representation of a Function

48 *Fundamentals of Measure Theory*

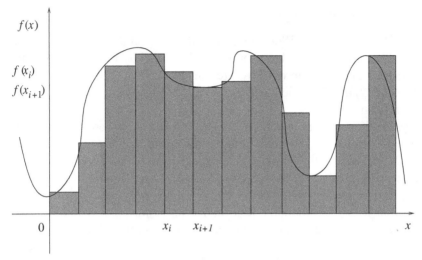

Figure 3.1 How to compute the Riemann integral.

by Means of a Trigonometrical Series", in which he extended the Cauchy integral to a larger class of functions, and established the necessary and sufficient conditions to represent a function by a Fourier series (Phillips, 1984).

The Cauchy–Riemann integral has the necessary features of a measure, and can be used to compute areas and volumes of functions that are common in Engineering or Mathematics. But it has some limitations, as can be seen in the following. Consider $(\mathbb{R}, \mathcal{B}(\mathbb{R}))$, the real line and its Borel set.

To compute the Riemann integral, as illustrated in Figure 3.1, the region under the curve, in the interval $[x_1, x_N]$, is divided into rectangles that have bases $\Delta x_i = x_{i+1} - x_i$ and heights $f(x_i)$, such that the total area between the points x_1 and x_N is approximated by

$$\text{Area} \approx \sum_{i=1}^{N} f_i \Delta x_i. \tag{3.9}$$

This method is similar to that used by the measurement officials in the ancient Egypt, who used ropes, marked with knots, to measure the lengths of the sides of rectangles, in order to determine the land area. The individual areas were added to obtain the total area. Of course, the approximation improves as the rectangles' bases decrease, to better fit the curve.

3.3 The Riemann Integral

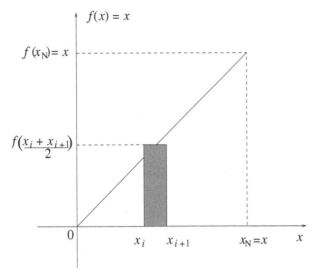

Figure 3.2 Computation of the Riemann integral, for the identity function.

Taking the limit of the sum, when the rectangle base goes to zero, one obtains, for $x_1 = a$ and $x_N = b$, the Riemann integral

$$I = \lim_{\Delta x_i \to 0} \sum_{i=1}^{N} f_i \Delta x_i = \int_a^b f(x) \, dx. \quad (3.10)$$

Example: The computation of the integral by the previous method in usually cumbersome. Consider the calculation of the area under the function $f(x) = x$, between the points $x_0 = 0$ and $x_N = x$, as shown in Figure 3.2.

Consider the use of the middle point to determine the rectangle's height to compute the area. By definition,

$$\int_0^x f(x) \, dx = \lim_{\Delta x_i \to 0} \sum_{i=0}^{N-1} f_i \Delta x_i$$

$$= \lim_{x_{i+1} \to x_i} \sum_{i=0}^{N-1} f\left(\frac{x_i + x_{i+1}}{2}\right) \cdot (x_{i+1} - x_i)$$

$$= \lim_{x_{i+1} \to x_i} \sum_{i=0}^{N-1} \left(\frac{x_i + x_{i+1}}{2}\right) \cdot (x_{i+1} - x_i)$$

$$= \lim_{x_{i+1} \to x_i} \frac{1}{2} \sum_{i=0}^{N-1} \left(x_{i+1}^2 - x_i^2\right)$$

$$= \lim_{x_{i+1} \to x_i} \frac{1}{2} \left[\left(x_N^2 - x_{N-1}^2\right) + \left(x_{N-1}^2 - x_{N-2}^2\right) + \cdots \right.$$
$$\left. + \left(x_2^2 - x_1^2\right) + \left(x_1^2 - x_0^2\right)\right]$$

$$= \frac{x_N^2 - x_0^2}{2} = \frac{x_N^2}{2},$$

substituting $x_N = x$, one obtains the known result,

$$\int_0^x f(x)\mathrm{d}x = \frac{x^2}{2}. \blacksquare$$

The Riemann integral is useful for most practical applications, including Engineering problems, because the functions are usually continuous and have primitives. But it lacks certain properties, related to set operations, that are needed to compute more complex functions of real variables (Wilcox and Myers, 1994).

Example: The Riemann integral can fail in those cases that involve limits of series of functions. The following function

$$f_n(x) = nxe^{-nx^2}u(x)$$

has the integral of the limit given by

$$\int_0^1 \lim_{n \to \infty} f_n(x)\mathrm{d}x = 0,$$

because $\lim_{n \to \infty} f_n(x) = 0$ for any x.

But,

$$\int_0^1 f_n(x)\mathrm{d}x = \int_0^1 nxe^{-nx^2}u(x)\mathrm{d}x = \frac{1 - e^{-n}}{2}.$$

Therefore, the limit of the integral is

$$\lim_{n \to \infty} \left(\int_0^1 f_n(x)\mathrm{d}x\right) = \lim_{n \to \infty} \left(\frac{1 - e^{-n}}{2}\right) = \frac{1}{2},$$

which is different from the previous result, obtained for the integral of the limit. $\blacksquare$

3.3 The Riemann Integral

Gaston Darboux (1842–1917), a French mathematician known for his work on differential geometry, simplified the formal computation of the Riemann integral, creating the inferior and superior sums for the function to be dealt with (Bressoud, 2008). However, the notions of *infimum* and *supremum*, that led to the concepts of Riemann superior and inferior integrals, were initially proposed by Vito Volterra (1860–1940), an Italian mathematician, in 1881.

For a function $f : [a, b] \longrightarrow \mathbb{R}$, which is real and limited, and for real numbers a, b, such that $a < b$, he defined the area as the limit of the approximation of circumscribed rectangles on the curve. The rectangle tops are M_i,

$$M_i = \sup_{x_i \leq x \leq x_{i+1}} f(x) \tag{3.11}$$

always above the curve, as illustrated in Figure 3.3, and inscribed rectangles on the curve, with heights, m_i,

$$m_i = \inf_{x_i \leq x \leq x_{i+1}} f(x) \tag{3.12}$$

which are below the curve, for $i \leq N$, as shown in Figure 3.4.

The upper sum (US) is defined for the approximation using circumscribed rectangles,

$$\text{US} \approx \sum_{i=1}^{N} M_i \Delta x_i, \tag{3.13}$$

in which the partition of $[a, b]$ forms a finite set $Q = \{x_1, x_2, \ldots, x_N\}$, with

$$a = x_1 < x_2 < x_3 < \cdots < x_N = b.$$

The lower sum (LS) is defined for the approximation using inscribed rectangles,

$$\text{LS} \approx \sum_{i=1}^{N} m_i \Delta x_i. \tag{3.14}$$

The Riemann integral exists when the lower sum Darboux limit

$$I_m = \lim_{\Delta x_i \to 0} \sum_{i=1}^{N} m_i \Delta x_i \tag{3.15}$$

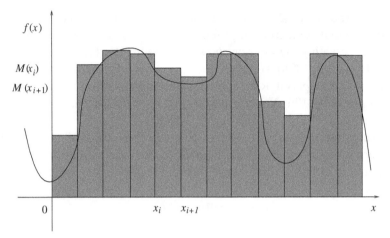

Figure 3.3 Sketch of the Riemann integral, upper limit.

and the upper sum Darboux limit

$$I_M = \lim_{\Delta x_i \to 0} \sum_{i=1}^{N} M_i \Delta x_i \qquad (3.16)$$

exist and are equal.

In order to guarantee that the difference between the areas is small, it is necessary to control the ripple amplitude, $M_i - m_i$, of the function $f(x)$ in the interval Δx_i. This can be done by taking small segments $\Delta x_i < \epsilon$, such that $M_i - m_i < \delta/(b-a)$. Therefore,

$$\sum_{i=1}^{N}(M_i - m_i)\Delta x_i < \frac{\delta}{(b-a)} \sum_{i=1}^{N} \Delta x_i = \frac{\delta}{b-a} \cdot (b-a) = \delta,$$

and that leads to the concept that every continuous function is integrable.

On the other hand, the Riemann integral properties are insufficient to define a measure that could be used for the whole family of Borel real line segments. This can be verified, if one considers a function $f(x)$ which assumes a constant value inside an interval, and zero outside the same interval.

Consider the indicator function, defined for the set of real numbers $\mathbb{R}$, which is known as Dirichlet function, in honor of the French mathematician Lejeune Dirichlet (1805–1859), defined as

$$I_\mathbb{Q}(x) = 1,$$

3.4 The Lebesgue Integral

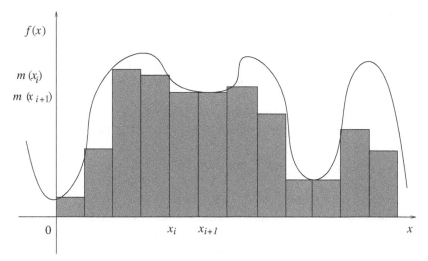

Figure 3.4 The lower limit for the Cauchy integral.

that is, it is equal to one for every number in the set of rationals, and

$$I_{\overline{\mathbb{Q}}}(x) = 0,$$

for the irrational numbers. This means that the Dirichlet function points to the rational numbers.

The Riemann integral is not well defined in this case, although the set $\mathbb{Q}$ is dense, because it has holes, it is denumerable. Its complement $\overline{\mathbb{Q}} = \mathbb{R} - \mathbb{Q}$ represents the set of irrational numbers, which has cardinality $\aleph$, the same as the set of real numbers.

Therefore, it is not clear what the limit $\Delta x_i \to 0$ really means in this case, because, for every chosen number, there is always a rational and an irrational number in the vicinity. Even if the Cauchy integral is used, the computation is not possible, because the upper sum is 1, while the lower sum is 0.

The Riemann integral is used whenever it is possible, that is, for functions that are calculated over finite unions of intervals. The more complex areas are obtained from limits of unions of intervals, if the limit makes sense.

But, for the case of σ-algebras, such as the Borel field, it is not possible to assume that the limit of the Riemann integral is equal to the integral of the limit. That is, the limit of an infinite sequence of convergent functions may be different from the original function's integral.

3.4 The Lebesgue Integral

Henri Léon Lebesgue (1875–1941) has found the solution for the computation of the area under the curve of complex functions. He realized that the Riemann integral was inappropriate as a measure for the Borel family of real line segments.

He proposed to group the domain subsets that were mapped into the same image, when he analyzed a strange function found by Dirichlet

$$\chi(x) = \lim_{m \to \infty} \left[\lim_{n \to \infty} (\cos m! \pi x)^{2n} \right], \quad (3.17)$$

that is discontinuous for every point in the domain, because it is 0 for all rational numbers and equal to 1 for the rational ones (Lebesgue, 1904).

Lebesgue compared his idea to the calculation of the statistical mean of a random variable, a really ingenuous insight, that had the objective of eliminating some of the deficiencies of the Riemann integral. The partition involved the image of the function, instead of the domain, as proposed by Riemann. Of course, it is necessary to associate a certain measure to the subsets of the real line, which can pose some difficulty (Wilcox and Myers, 1994).

Another improvement of the method is an increase in the abstraction level of the integral, because the Lebesgue measure may not be directly associated to the actual length of a certain segment, or rectangle. It can be related to the measurement of abstract entities, such as probability, that is, a given segment, rectangle or hyper-rectangle, may have a Lebesgue probability measure.

There are several graphical interpretations of the Lebesgue integral, and also some mathematical interpretations of it, which is one of the difficulties in dealing with the Lebesgue measure and integral.

Some authors interpret the integral as the sum of the areas limited by horizontal slices of the function, each one with width $\Delta f_i = f_{i+1} - f_i$, in place of the usual vertical rectangles, of width Δx_i, that compose the Riemann integral. Those rectangles are associated to the values x_i on the abscissa, as shown in Figure 3.5.

That interpretation is also related to the Stieltjes integral, which is an interesting manner of visualizing the computation of the area under the curve of a given function $f(x)$. The name of the integral honors Thomas Joannes Stieltjes (1856–1894), a Dutch mathematical who did not live long enough to see the fruits of his accomplishment.

3.4 The Lebesgue Integral

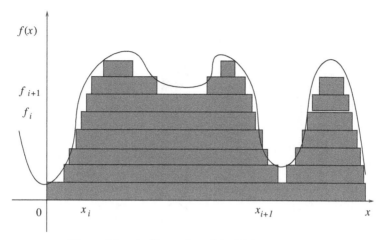

Figure 3.5 An illustration of the Stieltjes integral.

The Stieltjes integral of a function f in the interval $[a, b]$ can be written as

$$I = \int_a^b dF = \int_a^b dF(x), \tag{3.18}$$

in which $dF(x)$ represents the function increment.

The Stieltjes definite integral of f in terms of the generating function, or integral core P, in the interval $[a, b]$, is

$$I = \int_a^b f(x) dP(x). \tag{3.19}$$

The integral is useful for the calculation of functions of a random variable, such as the moments, in which P is a probability measure. If the core, or cumulative function, has a derivative, then the Stieltjes integral becomes the Riemann integral,

$$I = \int_a^b f(x) p(x) \mathrm{d}x, \tag{3.20}$$

in which $[a, b]$ is the interval of integration for the probability density function $p(x)$.

A measure, $\mu(A_i)$, is related to each set A_i, in the domain of the function, and it weighs the importance of the chosen points in the computation of the area. An illustration of the Lebesgue integral is shown in Figure 3.6.

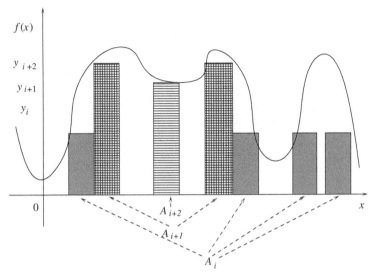

Figure 3.6 An illustration of the Lebesgue integral.

Each subset A_i is formed by joining together certain regions of the domain A that map into the same image in the range B, as defined by the function $y = f(x)$, and illustrated in Figure 3.7.

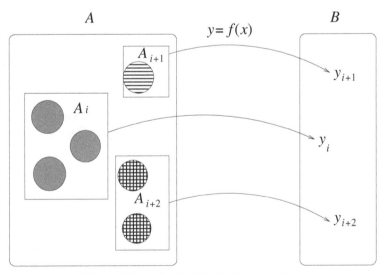

Figure 3.7 A sketch of the Lebesgue method.

Lebesgue expanded the area concept and managed to show that the integral could be used to compute the length of any rectifiable curve (Phillips, 1984). An adequate measure is needed to establish the necessary conditions for the existence of the Lebesgue integral. The measure must be a logical choice, and present a set of strict mathematical properties that could be exploited in practical situations, and also present a valid physical interpretation.

3.4.1 The Lebesgue Measure

The Lebesgue measure generalizes the concept of the length of an interval (a, b), that can be usually defined as $\lambda(a, b) = |b - a|$. This concept is extended to the family of Borel sets, which is a class of measurable sets, under the measure λ.

It is important to mention that the measure is not discrete, that is, $\lambda(\{x\}) = 0, \forall x \in \mathbb{R}$. As a consequence, every denumerable set has null Lebesgue measure. Of course, thare are also non-enumerable sets that present null Lebesgue measure. The Cantor set is such an example.

A measure is established for a given algebra of sets, that is, a family of sets that possess certain properties, and the measure is usually positive or zero, that is $\lambda \geq 0$. Furthermore, the measure is σ-additive,

$$\lambda\left(\bigcup_{i=1}^{\infty} F_i\right) = \sum_{i=1}^{\infty} \lambda(F_i), \tag{3.21}$$

for a family of measurable disjoint sets $\mathcal{F} = \{F_i\}$.

A measure space is a pair $(\mathcal{F}, \mu)$, for which, $\mathcal{F}$ is a σ-field, defined for a set X and $\mu : \mathcal{F} \mapsto [0, \infty]$ is a σ-additive function. The elements of $\mathcal{F}$ are said to be measurable and the function μ is a measure for X.

It is important to mention that the Lebesgue measure can be defined for $\mathbb{R}^n$, and, in this case, the rectangles are measurable sets whose measure coincides with their n-dimensional volume. Those sets are also called hyper-rectangles, n-dimensional rectangles, or n-dimensional segments.

Consider a set $A \subset \mathbb{R}^n$, the Euclidean n-dimensional space, and the open rectangles $R_1, R_2, \ldots$, that cover A. The Lebesgue exterior measure of the set A is defined as

$$\mu(A) = \inf \sum_{n=1}^{\infty} v(R_n), \tag{3.22}$$

in which $v(R_n)$ is the volume of the n-dimensional rectangle $R_n \subset \mathbb{R}$.

The Lebesgue exterior measure of the set A is defined for a σ-algebra $\mathcal{B}(\mathbb{R}^n)$, composed of all Borel subsets of $\mathbb{R}^n$, and satisfies the following properties:

1. $\mu(R) = v(R)$, if $R \subset \mathbb{R}^n$ is a rectangle, this is, the measure of a rectangle, or hyper-rectangle, is its n-dimensional volume.

    The measure for one dimensional segments is the length. The bi-dimensional measure is the area, and for three dimensions the measure is the volume.

2. $\mu(\emptyset) = 0$, that is, the measure of an empty set is null.

    Because $A \in \mathcal{B}(\mathbb{R}^n)$, then $\mu(A \cup \emptyset) = \mu(A) \cup \mu(\emptyset)$; therefore, $\mu(\emptyset) = 0$.

3. $\mu(A) \leq \mu(B)$, if $A \subset B$, that is, the measure is a monotonically increasing function.

    It is possible to put $B = (B - A) \cup A$, because of the disjunction, $\mu(B) = \mu(B - A) + \mu(A) \geq \mu(A)$.

4. $\mu(\{a\} \cup A) = \mu(A)$, if $a \in \mathbb{R}^n$.

    The Lebesgue measure is invariant to the addition of a point to the set. The union with a singleton does not increase the measure.

5. $\mu(A) = 0$, if and only if A is a zero measure set.

    It is interesting to note that zero measure sets are not necessarily empty. There are infinite sets with null measure.

6. $|\mu(A) - \mu(B)| \leq \mu(A \triangle B)$, for $A, B \subset \mathbb{R}^n$, with $\mu(A) < \infty$ or $\mu(B) < \infty$.

    From the property of sets, $A \triangle B = (A - B) \cup (B - A)$; therefore, by disjunction, $\mu(A \triangle B) = \mu(A - B) + \mu(B - A) \geq |\mu(A) - \mu(B)|$. Equality occurs if $A \subset B$ or $B \subset A$.

7. $\mu(A) \leq \sum_{n=1}^{\infty} \mu(A_n)$, if $A = \bigcup_{n=1}^{\infty} A_n$, that is, it is a sub-additive function.

    Equality occurs if the sets A_n are mutually disjoint.

Example: The identity function is defined as $\mu(x) = \mu[0, x]$, $x \in \mathbb{R}$. This function has the necessary features for a measure. If μ is a positive line, extending from the origin, $\mu(x)$ represents its length. Of course, $\mu(0) = \mu[0,0] = 0$, and $\mu(b) - \mu(a) = \mu[a,b] = b - a$, if $a \leq b$. ∎

A set $A \subset \mathbb{R}$ is Lebesgue measurable if, for all $\epsilon > 0$, there are rectangles $R_1, R_2, \ldots$ such that their countable union

$$U = \sum_{n=1}^{\infty} R_n$$

satisfies the following condition for the symmetric difference measure

$$\mu(A \, \Delta \, U) < \epsilon,$$

This is illustrated in Figure 3.8. In other words, the symmetric difference between the set A and the denumerable union of rectangles, that results in U, can be considered negligible. The geometric interpretation of the property is that a set is measurable if it can be approximated by a countable union of rectangles, and it is not necessary that the rectangles be disjoint.

For the sets $A, B \in \mathcal{B}(\mathbb{R}^n)$, with measure $\mu(A) < \infty$ or $\mu(B) < \infty$, one has

$$|\mu(A) - \mu(B)| \leq \mu(A \, \Delta \, U). \tag{3.23}$$

Example: The Cantor set is constructed using a recursive subtraction of the central part of a line segment of unit length. Consider that the initial

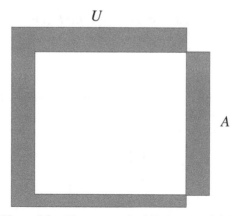

Figure 3.8 The symmetric difference set $A \, \Delta \, U$.

segment is the set $A_1 \subset U$, and form the sequence of sets:

$$A_2 = A_1 - \left(\frac{1}{2}, \frac{2}{3}\right),$$

$$A_3 = A_2 - \left[\left(\frac{1}{9}, \frac{2}{9}\right) \cup \left(\frac{7}{9}, \frac{8}{9}\right)\right],$$

$$\vdots$$

$$A_n = A_{n-1} - \left[\left(\frac{1}{3}, \frac{2}{3^{n-1}}\right) \cup \cdots \cup \left(\frac{3^{n-1}-2}{3^{n-1}}, \frac{3^{n-1}-1}{3^{n-1}}\right)\right].$$

The Cantor set, defined as

$$A = \bigcap_{n=1}^{\infty} A_n,$$

is non-denumerable and measurable, but has zero Lebesgue measure. ∎

To demonstrate that A is uncountable, it is only necessary to expand the numbers of U in a ternary base. In this case, A is the set of all points in U which do not have 1 in their ternary expansion, that is, the set is non-denumerable.

The limit superior, or *limes supremum*, of a sequence of sets $B_1, B_2, B_3, \ldots$, that belong to a field $\mathcal{F}$, is defined as

$$\limsup B_n = \bigcap_{k=1}^{\infty} \bigcup_{n \geq k} B_n, \qquad (3.24)$$

also called B_i, infinitely frequent. Let $B = \limsup B_n$, if

$$\sum_{k=1}^{\infty} \mu(B_k) < \infty,$$

then $\mu(B) = 0$. The result is known as the First Borel–Cantelli Lemma (Adams and Guillemin, 1996).

In order to demonstrate the lemma, let

$$A_k = \bigcup_{n \geq k} B_n,$$

such that,

$$B = \bigcap_{k=1}^{\infty} A_k.$$

As a particular case, $B \subset A_k$ for all k. By the sub-additivity property,

$$\mu(A_k) \leq \sum_{n \geq k} \mu(B_n).$$

Therefore, considering that

$$\sum_{k=1}^{\infty} \mu(B_k) < \infty,$$

for $\epsilon > 0$ there is a $k > 0$ such that

$$\mu(A_k) \leq \sum_{n \geq k} \mu(B_n) < \epsilon.$$

Because $B \subset A_k$, one has $\mu(B) < \epsilon$, but, considering that ϵ is arbitrary, one must have $\mu(B) = 0$.

3.4.2 Concept of the Lebesgue Integral

The Lebesgue integral is introduced in this section, based on the measure concept discussed previously. First, consider the function $f : A \mapsto \mathbb{R}$, in which $A = [a, b] \subset \mathbb{R}$. To compute the Lebesgue integral, it is necessary to determine the preimage of each value assumed by the function. This value is multiplied by the measure, which could be a length, an area, or a volume, of the preimage, and the result is added.

$$\int_A f d\mu = \sum_{k=1}^{M} y_k \mu(A_k). \tag{3.25}$$

The Lebesgue integration requires that the function be measurable. A function $f : A \mapsto \mathbb{R}$, defined on a measurable set $A \subset \mathbb{R}^n$, is measurable, if the set

$$f^{-1}(c, \infty) = \{x \in A : f(x) > c\}, \tag{3.26}$$

is measurable for every constant $c \in \mathbb{R}$.

Example: If a function $f : \mathbb{R} \mapsto \mathbb{R}$ is continuous, then it is measurable.

Because the set (c, ∞) is open, and f is continuous, then $f^{-1}(c, \infty)$ is open, therefore, measurable. Recall that every open set is a countable union of rectangles. ∎

Example: A function f is measurable, if it is the limit of a sequence of step functions everywhere. The sequence of step functions

$$g_n(x) = \begin{cases} \frac{n}{k}, & \text{if } \frac{(k-1)}{n} < x \leq \frac{k}{n},\ k = 1, 2, \ldots, n, \\ 0, & \text{otherwise}, \end{cases}$$

increases monotonically to the function

$$f_n(x) = \begin{cases} \frac{1}{x}, & \text{if } 0 < x \leq 1, \\ 0, & \text{otherwise}. \end{cases}$$

Then, f is measurable. But, the sequence of integrals

$$h_n = \int g_n = 1 + \frac{1}{2} + \frac{1}{3} + \cdots + \frac{1}{n}$$

does not converge, and f is not integrable. ∎

3.4.3 Properties of the Lebesgue Integral

For a function $f : A \mapsto \mathbb{R}$, defined on a measurable set $A \subset \mathbb{R}^n$, the following statements are equivalent:

1. $\{x \in A : f(x) > c\}$, is measurable;
2. $\{x \in A : f(x) \geq c\}$, is measurable;
3. $\{x \in A : f(x) < c\}$, is measurable;
4. $\{x \in A : f(x) \leq c\}$, is measurable.

The propositions can be proved, if one considers that

1. $\{x \in A : f(x) > c\} = A - \{x \in A : f(x) \leq c\}$;
2. $\{x \in A : f(x) \geq c\} = \bigcap_{n=1}^{\infty} \{x \in A : f(x) > c - \frac{1}{n}\}$;
3. $\{x \in A : f(x) < c\} = A - \{x \in A : f(x) \geq c\}$;
4. $\{x \in A : f(x) \leq c\} = \bigcap_{n=1}^{\infty} \{x \in A : f(x) < c + \frac{1}{n}\}$.

Example: The modulus function $|f| : \mathbb{R} \mapsto [0, \infty)$ is measurable. To demonstrate this, just verify that $\{x \in A : |f(x)| > c\} = \{x \in A : f(x) > c\} \cup \{x \in A : f(x) < -c\}$, which is a union of measurable functions. ∎

A simple function, $s : \mathbb{R}^n \mapsto \mathbb{R}$, has a finite image, that is, it assumes a finite number of values. The constant function is simple, and therefore, measurable. Let $f : \Omega \mapsto \mathbb{R}$, $f(\omega) = c$, then the set

$$\{x : -\infty < f(x) \leq c\} = \begin{cases} \emptyset, & \text{if } x < c, \\ \Omega, & \text{if } x \geq c. \end{cases}$$

Therefore, f is measurable, because

$$\{x : -\infty < f(x) \leq c\} \in \mathcal{F}.$$

The indicator function for $A \subset \mathbb{R}^n$, defined in the following, is simple.

$$I_A(x) = \begin{cases} 1, & \text{for } x \in A \\ 0, & \text{for } x \notin A. \end{cases}$$

Every simple function $s : \mathbb{R}^n \mapsto \mathbb{R}$ is a linear combination of indicator functions. To prove this, it is only necessary to note that the image of f is the set $\{c_1, c_2, \ldots, c_M\}$, and consider that

$$A_i = \{x \in \mathbb{R}^n : f(x) = c_i\},$$

such that

$$s = \sum_{i=1}^{M} c_i I_{A_i}. \tag{3.27}$$

A simple function s is measurable, if and only if, the sets A_i are measurable. Any function can be approximated by simple functions, if the coefficients are properly chosen.

A succession of simple functions is defined as $\{f_k\}$, $k \in \mathbb{N}$, that converge to $f : A \mapsto \mathbb{R}$

$$\lim_{k \to \infty} s_k(x) = f(x), \forall x \in A.$$

Example: The following functions $\sup f_n$ (supremum), $\inf f_n$ (infimum), $\lim_{n \to \infty} \sup f_n$ (limit superior) and $\lim_{n \to \infty} \inf f_n$ (limit inferior) are measurable.

Consider the supremum function $g(x) = \sup f_n$. One has,

$$\{x \in A : g(x) > c\} = \bigcup_{n=1}^{\infty} \{x \in A : f_n(x) > c\},$$

that is, $\sup f_n$ is measurable. It is possible to use the same argument to show that $\inf f_n$ is measurable.

64 Fundamentals of Measure Theory

Following the same logic, $\lim_{n\to\infty} \sup f_n = \inf g_m$, in which $g_m(x) = \sup\{f_n(x) : n \geq m\}$, and one notices that $\lim_{n\to\infty} \sup f_n$ is measurable. By analogy, $\lim_{n\to\infty} \inf f_n$ is also measurable. ∎

Therefore, the limit of sequences of measurable functions are also measurable. But, it is important to note that a composition of functions is not, necessarily, a measurable function. Even the composition $f(g(x))$, in which f is measurable and g is continuous, can result in a non measurable function.

Example: The quantizer is a device that maps a continuous signal into a discrete one (Alencar, 2011). The quantizer function is simple,

$$s = \sum_{i=-M}^{M} c_i I_{A_i} \qquad (3.28)$$

in which $A_i = (x_i, x_{i+1})$, and $\mu(A_i) = x_{i+1} - x_i$. ∎

Figure 3.9 illustrates the quantization process, for a constant step uniform quantizer.

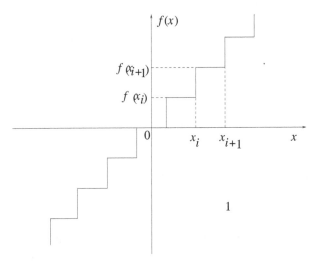

Figure 3.9 Quantization process.

Let $f : A \mapsto \mathbb{R}$ be a non negative measurable function, defined on a measurable set. The Lebesgue integral of f on A is defined as

$$\int_A f d\mu = \sup\{I_A(s) : 0 \leq s \leq f\}, \qquad (3.29)$$

if s is a simple function.

The Lebesgue integral of a simple function $s : \mathbb{R}^n \longrightarrow \mathbb{R}$ is

$$\int_A s d\mu = I_A(s).$$

From the definition of the Lebesgue integral for non-negative functions, it is possible to compute the integral of any measurable function, using the decomposition $f = f^+ + f^-$, in which the functions are defined as: $f^+ = \max(f, 0)$ and $f^- = -\min(f, 0)$. Therefore, for a measurable function $f : A \mapsto \mathbb{R}$, defined on a measurable set, the Lebesgue integral is

$$\int_A f d\mu = \int_A f^+ d\mu - \int_A f^- d\mu, \tag{3.30}$$

given that one of the integrals on the right-hand side should be finite.

Consider a set $A \subset \mathcal{F}$. The Lebesgue integral of the indicator function I_A, that maps X into $\mathbb{R}$, relative to the measure μ, is

$$\int_X I_A d\mu = \mu(A). \tag{3.31}$$

As a consequence, for any $\alpha \in \mathbb{R}$, the Lebesgue integral of αI_A, considered as a function of X in $\mathbb{R}$, relative to the measure μ, is given by

$$\int_X \alpha I_A d\mu = \alpha \cdot \mu(A), \tag{3.32}$$

with the convention $\infty \cdot 0 = -\infty \cdot 0 = 0$.

It can be demonstrated that the Lebesgue integral is an extension of the Riemann integral, that is, if $f : A \mapsto \mathbb{R}$ is a Riemann integrable functions, then it is also Lebesgue integrable, and the integrals are equal. In addition, the Lebesgue integral extends the concept to a broader class of functions.

The limit of integrable functions is not necessarily Riemann integrable itself, even for continuous functions; therefore the Lebesgue integral is useful to solve that type of problem, which is not infrequent when dealing with Calculus.

Example: The Dirichlet function can be integrated, in a certain interval, using the Lebesgue integral.

The measure of the closed interval $[0, 1]$, on the real line, is $\mu[0, 1] = 1$. Considering that the set of rationals $\mathbb{Q}$ is countable, its Lebesgue measure is zero in the interval. Therefore, the Lebesgue measure of the set of irrationals $\mu([0, 1] - \mathbb{Q} \cap [0, 1]) = 1$. ■

The Lebesgue theory is more comprehensive, and can be used to compute the integral of set functions and limits of functions. It guarantees a theoretical framework to support the concepts of measure and the integral, and can be used to build abstract measure spaces, such as the probability spaces.

Example: Compute the integral of the function $f : [0,1] \mapsto \mathbb{R}$, defined as $f(x) = 0$ in the Cantor set, and $f(x) = n$ for all x in each interval of length 3^{-n} which has been removed from the original interval $[0,1]$.

The function is given as,

$$f(x) = \sum_{n=1}^{\infty} n 2^{n-1} I_{A_n}(x),$$

in which A_n is the union of 2^{n-1} intervals, each one of length 3^{-n}, that are removed from the interval $[0,1]$ at the n-th step, and I_{A_n} is the indicator function of this set.

Because the convergence is monotonic,

$$\int_{[0,1]} f \, d\mu = \lim_{k \to \infty} \sum_{n=1}^{k} n \frac{2^{n-1}}{3^n} = \frac{1}{3} \sum_{n=1}^{\infty} n \left(\frac{2}{3}\right)^{n-1} = \frac{1}{3} \frac{1}{(1-\frac{2}{3})^2} = 3. \blacksquare$$

It is important to state again that any Riemann integrable function, $f : [a,b] \mapsto \mathbb{R}$, is also Lebesgue integrable, and the results are the same.

3.5 Henri Lebesgue

Henri Léon Lebesgue's theory of integration was published originally in his dissertation *Intégrale, Longueur, Aire* (Integral, Length, Area), at the University of Nancy, in 1902. The work appeared in the *Annali di Matematica*, in the same year. Félix Édouard Justin Émile Borel (1871–1956), a French mathematician and politician, best known for his founding work in the areas of measure theory and probability, was Lebesgue's supervisor. A picture of Lebesgue is shown in Figure 3.10.

Lebesgue formulated the theory of measure, in 1901, in his famous paper *Sur une Généralisation de l'Intégrale Définie* (On a Generalization of the Definite Integral), which appeared in the Comptes Rendus, in 1901. He proposed the definition of the Lebesgue integral, that generalises the notion of the Riemann integral and extends the concept of area below a curve to include discontinuous functions.

Lebesgue used to say that his integral was equivalent to computing the statistical average of a random variable using the formula. First, one counts

Figure 3.10 A portrait of Henri Léon Lebesgue. Adapted from: Public Domain, https://commons.wikimedia. org/w/index.php?curid=336482

how many outputs are favorable to certain values, and then use the relative frequency to obtain the mean.

The result is the same that is obtained if all the values are added and then the result is divided by the number of values. According to Lebesgue, the discrepancy between the results appears when the set is infinite and has infinite discontinuities.

If the function has an area, or integral, in the Riemann sense, the result is equal to that obtained with the Lebesgue integral. But, there are functions, such as Dirichlet's, which are not measurable using Riemann's method, but can be integrated using Lebesgue's.

3.6 Problems

1. Consider that the measure $\mu(B)$ is finite and nonzero, and demonstrate that $\nu_B(A)$ defines a probability measure.

$$\nu_B(A) = \frac{\mu_B(A)}{\mu(B)} = \frac{\mu(A \cap B)}{\mu(B)}, \forall A, B \in \mathcal{F}.$$

2. Give an example of a function for which the Riemann integral does not make sense.

3. Explain why the Lebesgue measure generalizes the concept of the length of an interval (a, b).

4. Considering that the length of an interval can be usually defined as $\lambda(a, b) = |b - a|$, extend the concept to the family of Borel sets in two dimensions.

5. Show that the Lebesgue measure is invariant to the addition of a point to the set, that is, the union with a singleton does not increase the measure,

$$\mu(\{a\} \cup A) = \mu(A),$$

if $a \in \mathbb{R}^n$.

4

Generalized Functions

"For me, everything in nature is mathematics."
René Descartes

4.1 A Note on Generalized Functions

Probability theory sometimes deals with distribution, or generalized, functions, which are important mathematical constructs that permit the computation of integrals of functions that describe important, but sometimes strange, phenomena. A generalized function is a mathematical concept that generalizes the usual notion of a function. This generalization is necessary in many problems in probability, engineering, physics, mathematics and other areas of science.

Using the generalized functions it is possible to express idealized concepts, such as the probability density of a single event on the real line, the density of a material point, such as a black hole, a point charge or a point dipole, or the intensity of an instantaneous source in a mathematically manageable form. The use of generalized functions also permits to deal with the Riemann integral, formulated by Georg Friedrich Bernhard Riemann (1826–1866), a German mathematician who made important contributions to analysis, number theory, and differential geometry.

The concept of generalized function reflects the physical fact that a real quantity cannot be measured at a point, but that only the mean value over a sufficiently small neighbourhood of a given point can be measured. Therefore, the technique of generalized functions is a convenient way to describe the distribution of physical quantities, and because of this characteristic the generalized functions are also called distributions.

The British electrical engineer and theoretical physicist Paul Adrien Maurice Dirac (1902–1984) introduced the generalized functions in his seminal book "The Principles of Quantum Mechanics," and used the concept in his research on quantum mechanics. But, the delta function had been used by Oliver Heaviside (1850–1925), an English mathematician and physicist, in his operational calculus long before it was widely accepted in the mathematical community.

Dirac made fundamental contributions to the development of quantum mechanics and quantum electrodynamics, such as the prediction of the existence of antiparticles. He also made systematic use of the concept of the delta function and its derivatives, and contributed to mediate the general relativity and the quantum mechanics theories. Dirac shared the 1933 Nobel Prize in Physics with Erwin Rudolf Josef Alexander Schrödinger (1887–1961), an Austrian theoretical physicist.

The foundations of the mathematical theory of generalized functions were laid by Sergei Lvovich Sobolev (1908–1989), a Soviet mathematician, in 1936, who solved the problem for hyperbolic equations, stated by Augustin Louis Cauchy (1789–1857), and by Laurent-Moïse Schwartz (1915–2002), a French mathematician, who gave a systematic account of the theory of generalized functions, indicated many applications, and received the Fields Medal for his elaboration on the theory of distributions. Their work transformed and generalized the classical theory of functions, to produce the modern theory of distributions, or generalized functions.

4.2 The Unit Step Function

The introduction of the concept of unit step function by Heaviside is the usual point of departure to present the generalized functions, because it can be easily understood and has many important applications. The definition of a classical function implies determining a rule to describe how this function maps one set of real numbers to another set, that is, $f : \mathbb{R} \to \mathbb{R}$.

By this definition, a step function is a real function in the usual way. The problems arise, as shown in the next section, when it is necessary to take the derivative of a step function, because the unit step function has a jump at $x = 0$, and such type of discontinuity is usually avoided in the study of Differential Calculus. A way to circumvent this difficulty is to use the convenient concept of functional.

The unit step function is depicted in Figure 4.1, and is defined as

$$\mathrm{u}(x) = \begin{cases} 0 & \text{if } x \in (-\infty, 0) \\ 1 & \text{if } x \in (0, \infty). \end{cases} \qquad (4.1)$$

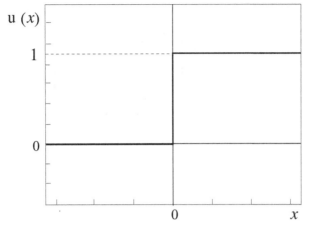

Figure 4.1 A unit step function.

The value of the unit step function at zero is usually undefined. Some authors prefer to define u(0) = 1/2, for reasons of expediency, others prefer to attribute the values zero or one, according to the particular application.

The unit step function spans the whole real line $\mathbb{R}$. It can indicate the exact point moment a certain phenomenon occurs, for instance, when a switch in closed, or a signal is turned on. It is also used to indicate the concept of belonging in Set Theory.

Example: The unit step function can be used to generate a digital waveform from a sinusoidal function, using a composite function. Consider that the input signal is $f(t) = \cos(\omega t)$. Then, the required formula for the output signal can be written as

$$y(t) = \mathrm{u}[f(t)] = \mathrm{u}[\cos(\omega t)] = \begin{cases} 1, & \text{if } f(t) \in (0, \infty) \\ 0, & \text{if } f(t) \in (-\infty, 0), \end{cases} \quad (4.2)$$

and the output signal is depicted in Figure 4.2. Note that the signal is a square wave, a periodic signal that can only assume the values 0 and 1. ∎

4.2.1 Properties of the Unit Step Function

The unit step function is a very useful one for applications in Mathematics and Engineering, because it can be used to delimit other functions. Some of its interesting properties are presented in the following.

$$\int_{-\infty}^{\infty} f(x)\mathrm{u}(x)\mathrm{d}x = \int_{0}^{\infty} f(x)\mathrm{d}x, \quad (4.3)$$

72 Generalized Functions

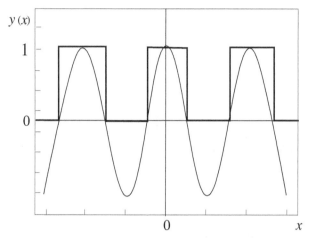

Figure 4.2 A square wave produced from a cosine wave, using the unit step function.

This illustrates a typical application of the unit step function, to delimit the range of integration of a given function.

Note, in the following property, that the unit step function has been inverted, regarding the abscissa, and shifted to the position t on the τ axis. Thus, $\mathrm{u}(x - \tau)$ is equal to one from $\tau = -\infty$ to $\tau = t$, and zero elsewhere.

$$\int_{-\infty}^{\infty} f(\tau)\mathrm{u}(x - \tau)\mathrm{d}\tau = \int_{-\infty}^{x} f(\tau)\mathrm{d}\tau. \tag{4.4}$$

4.3 The Signum Function

The signum, or sign, function attributes one to positive values of the independent variable, and minus one to the negative values, as follows

$$\mathrm{sgn}(x) = \begin{cases} 1, & \text{if } x \in (0, \infty) \\ -1, & \text{if } x \in (-\infty, 0). \end{cases} \tag{4.5}$$

The signum, or sign, function, depicted in Figure 4.3, is usually not defined for $x = 0$, but the value zero can be attributed in some cases to facilitate limiting operations.

The sign function can be defined with the help of the unit step function as

$$\mathrm{sgn}(x) = \mathrm{u}(x) - \mathrm{u}(-x), \tag{4.6}$$

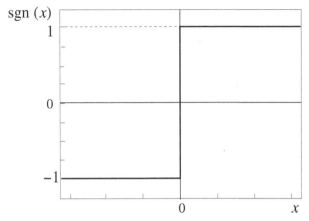

Figure 4.3 The signum function.

Example: A formula for an alternating square wave can be written using the sgn(x). It is only necessary to write $x(t) = \cos t$, and the composite function is $f(t) = \text{sgn}[\cos(t)]$. ∎

4.4 The Gate Function

The gate function, $g(x)$, plays an important role in the study of signals and systems, because it can be used to represent a pulse signal, the response of a linear system to a given signal, or the time a certain circuit is in operation, for example. Figure 4.4 illustrates the gate function.

The gate function can be obtained as the composition of two displaced unit step functions,

$$g(x) = \text{u}(x+a) - \text{u}(x-a), \qquad (4.7)$$

4.5 The Impulse Function

The impulse function, also called Dirac function, or Dirac delta distribution, after the Paul Dirac, is an important mathematical construct that helps explain several physical, mechanical, electrical and probabilistic phenomena. An impulse function can be seen in Figure 4.5.

The impulse function, that does not meet the requirements of a typical function, but can be considered as a generalized function, is defined as

74 *Generalized Functions*

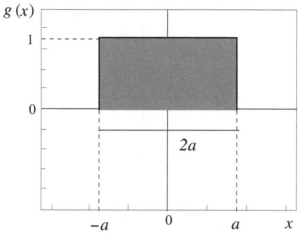

Figure 4.4 The Gate function.

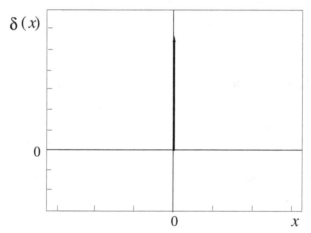

Figure 4.5 The impulse function.

follows,

$$\delta(x) = \begin{cases} \infty, & \text{if } x = 0 \\ 0, & \text{if } x \neq 0. \end{cases} \qquad (4.8)$$

The value of the impulse function at zero is infinity. This can cause mathematical problems, sometimes, and therefore some authors prefer to define the impulse function as a limit. For instance, it can be defined as the

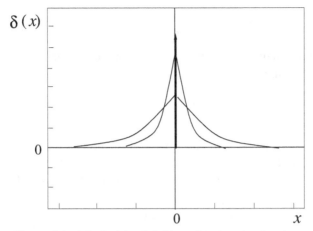

Figure 4.6 The limiting definition of the impulse function.

limit of the Laplace function as the parameter α goes to infinity,

$$\delta(x) = \lim_{\alpha \to \infty} \frac{\alpha}{2} e^{-\alpha |x|}, \tag{4.9}$$

This can be observed in Figure 4.6. Note that the Laplace function is squeezed, as $\alpha \to \infty$, but the area under the curve remains unitary.

The Laplace function was named after Pierre-Simon, marquis de Laplace (1749–1827), a French scholar and polymath with important contributions to the development of Engineering, Mathematics, Statistics, Physics, Astronomy, and Philosophy.

Note that the Laplace function is normalized, as can be verified, which means that it can be used as a probability distribution function,

$$\int_{-\infty}^{\infty} \frac{\alpha}{2} e^{-\alpha |x|} \mathrm{d}x = \int_{-\infty}^{0} \frac{\alpha}{2} e^{\alpha x} \mathrm{d}x + \int_{0}^{\infty} \frac{\alpha}{2} e^{-\alpha x} \mathrm{d}x$$
$$= \frac{\alpha}{2} \left[\frac{e^{\alpha x}}{\alpha} \right]_{-\infty}^{0} + \frac{\alpha}{2} \left[-\frac{e^{-\alpha x}}{\alpha} \right]_{0}^{\infty} = 1. \tag{4.10}$$

The same result for the impulse can also be obtained from the limit of a normalized gate function, considering the interval $(0, \tau)$, which one defines as $g_\tau(x)$,

$$g_\tau(x) = \frac{1}{\tau} [\mathrm{u}(x) - \mathrm{u}(x - \tau)]. \tag{4.11}$$

as the parameter $\tau \to 0$.

76 Generalized Functions

$$\delta(x) = \lim_{\tau \to 0} g_\tau(x) = \lim_{\tau \to 0} \frac{\mathrm{u}(x) - \mathrm{u}(x - \tau)}{\tau}. \tag{4.12}$$

The operational definition of the impulse function is therefore related to the unit step function, as follows,

$$\frac{\mathrm{du}(x)}{\mathrm{d}x} = \delta(x), \tag{4.13}$$

that is, the impulse function is the derivative of the unit step function. This is a useful relation, and several examples are given to emphasize its importance.

Therefore, because of the Fundamental Theorem of the Calculus,

$$\int_{-\infty}^{x} \delta(\tau)\mathrm{d}\tau = \mathrm{u}(x). \tag{4.14}$$

Example: The quantizer is a device that maps a continuous signal into a discrete one, which is given by Formula 3.28, that uses simple functions. It can be described with the help of the unit step function, as

$$q(x) = \sum_i \mathrm{u}(x - x_i),\ i \in \mathbb{N},\ x \geq 0, \tag{4.15}$$

and $-q(-x)$ for negative x. If one takes the derivative of $q(x)$, results in

$$\Delta(x) = q'(x) = \sum_i \delta(x - x_i),\ i \in \mathbb{Z},\ x \in \mathbb{R}, \tag{4.16}$$

which is an infinite sequence of impulses, located at x_i, as shown in Figure 4.7.∎

The impulse function has other important properties, for example,

$$\int_{-\infty}^{\infty} \delta(x)\mathrm{d}x = 1, \tag{4.17}$$

the area of the impulse function is one, which means that the impulse function is normalized.

Example: The derivative of the sign function can be computed using the properties of the unit step function.

$$\frac{\mathrm{d}}{\mathrm{d}x}\mathrm{sgn}(x) = \frac{\mathrm{d}}{\mathrm{d}x}[\mathrm{u}(x) - \mathrm{u}(-x)] = \delta(x) + \delta(-x) = 2\delta(x),$$

which is true, because the impulse is considered an even function.∎

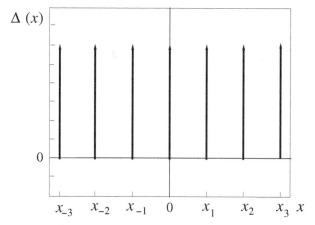

Figure 4.7 An infinite series of impulses.

The impulse function can also extract a sample of a given function, as can be observed in the following,

$$\int_{-\infty}^{\infty} f(x)\delta(x)\mathrm{d}x = f(0), \tag{4.18}$$

and as displayed in Figure 4.8.

$$\int_{-\infty}^{\infty} f(x)\delta(x-\tau)\mathrm{d}x = f(\tau). \tag{4.19}$$

The impulse function can be used to model voltage, current or power peaks. For example, a lightning can be modeled as a voltage impulse, because it occurs during a short time interval and has an amplitude that can reach millions of volts.

Example: A series of alternating impulses can be created based on the formula for the square wave previously determined, $y(t) = \mathrm{u}[f(t)] = \mathrm{u}[\cos(\omega t)]$.

The derivative of $y(t)$ can be obtained using the formula for the derivative of a composite function, that is,

$$z(t) = \frac{\mathrm{d}y(t)}{\mathrm{d}t} = \frac{\mathrm{d}u(t)}{\mathrm{d}f(t)} \cdot \frac{\mathrm{d}f(t)}{\mathrm{d}t}. \tag{4.20}$$

$$z(t) = \frac{\mathrm{d}\,\mathrm{u}[\cos(\omega t)]}{\mathrm{d}t} = \delta[\cos(\omega t)] \cdot [-\omega\sin(\omega t)]. \tag{4.21}$$

78 Generalized Functions

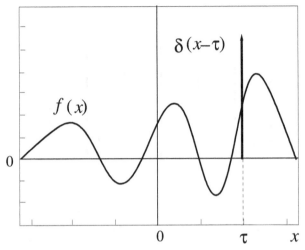

Figure 4.8 The impulse function shifted to τ to sample the function $f(x)$.

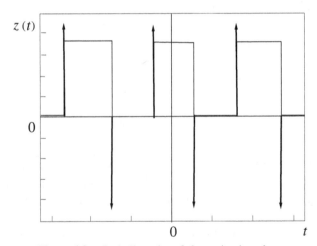

Figure 4.9 Periodic series of alternating impulses.

$$z(t) = -\omega\delta[\cos(\omega t)] \cdot \sin(\omega t), \qquad (4.22)$$

which is the formula for the series of alternating impulses depicted in Figure 4.9. ∎

4.5.1 The Functional

The mathematical difficulties produced with the use of objects such as the impulse, led to the creation of a functional, which is a function of functions. As a function is a unique mapping from one set of numbers to another, a functional $\mathcal{F}$ can be defined as a mapping $\mathcal{F} : \mathcal{A} \to \mathbb{R}$, in which $\mathcal{A}$ is a class of functions. Therefore, a functional maps functions to real numbers.

An example of a functional of this type is the Riemann definite integral, as follows

$$I_A[f] = \int_A f(x)\mathrm{d}x, \tag{4.23}$$

which takes a function $f(x)$ from a set of suitably integrable functions and maps it to a real number, that is the value of the integral, or the area under the curve $f(x)$ in the set A. The functional concept can be used to define distributions, and remove the difficulty which Dirac and other mathematicians faced when differentiating the impulse function.

Therefore, one defines the functional $\mathcal{F}[\phi]$ as

$$\mathcal{F}[\phi(x)] = \int_{-\infty}^{\infty} f(x)\phi(x)\mathrm{d}x, \tag{4.24}$$

in which $\phi(x)$ is a test function, the argument of the functional, an infinitely continuously differentiable function with compact support, and $f(x)$ is the kernel, which determines the properties of the functional.

Therefore, analogous to the way a function acts on an input number and produces an output, a specific distribution, $\mathcal{F}[\phi(x)]$, is defined by the manner in which it transforms test functions into numbers. The mentioned functions are real, and map points from $\mathbb{R}$ into $\mathbb{R}$.

The linear functional $\mathcal{F}[\phi(x)]$ maps the points of the function $\phi(x)$, of a certain functional space $\mathcal{A}$, called space of text function, onto points of the real line $\mathbb{R}$.

The Dirac delta distribution is only one of infinitely many distributions which do not correspond to classical functions. It is possible to obtain additional distributions by differentiating the impulse function in the sense of distributions. Consider generalizing the concept of differentiation to apply to distributions,

$$\mathcal{F}'[\phi(x)] = \int_{-\infty}^{\infty} f'(x)\phi(x)\mathrm{d}x, \tag{4.25}$$

in which the function f is continuously differentiable. Calculating using integration by parts, one obtains

$$\mathcal{F}'[\phi(x)] = [f(x)\phi(x)]_{-\infty}^{\infty} - \int_{-\infty}^{\infty} f(x)\phi'(x)\mathrm{d}x. \quad (4.26)$$

But the term in brackets vanishes since $\phi(x)$, as a test function, has compact support. Thus,

$$\mathcal{F}'[\phi(x)] = - \int_{-\infty}^{\infty} f(x)\phi'(x)\mathrm{d}x. \quad (4.27)$$

This last result is sometimes referred to as the weak derivative, since it extends derivatives to functions which normally would not be differentiable. This can be generalized to a great extent, to include higher order derivatives (Taylor, 2022).

A functional $\mathcal{F}[\phi(x)]$ is called linear, if, for all test functions $\phi(x)$ and $\varphi(x)$, the following equality holds

$$\mathcal{F}[\alpha\phi + \beta\varphi] = \alpha\mathcal{F}[\phi] + \beta[\varphi]. \quad (4.28)$$

in which α and β are arbitrary constants.

A functional F on $\mathcal{A}$ is called continuous, if for any functional sequence $\{\phi_k(x)\}$, whose elements belong to the space $\mathcal{A}$ and converge at $k \to \infty$ to a certain test function $\phi(x) \in \mathcal{A}$, the corresponding numerical sequence $\{\mathcal{F}[\phi_k]\}$ converges to the number $\mathcal{F}[\phi]$.

As observed, it is possible to add distributions, but it is not possible to multiply distributions when their singular support coincide. Despite that, it is possible to take the derivative of a distribution, to obtain other distributions (Todd, 2022).

Some linear continuous functionals $T[\phi]$ cannot be identified with a certain continuous kernel $f(x)$. In this case, this functional is dubbed a singular generalized function. The most useful example of a singular generalized function is the functional

$$\delta[\phi(x)] = \int_{\infty}^{\infty} \delta(x)\phi(x)\mathrm{d}x = \phi(0), \quad (4.29)$$

that maps the test function $\phi(x) \in \mathcal{A}$ onto its value at $x = 0$. This linear and continuous functional is called a delta function.

The delta, or impulse, function belongs to a narrow class of scale-invariant functions, whose argument may be dimensional. This means that it has a non-zero dimension that depends on the dimension of the argument. In other words, the impulse function, which has the spatial coordinate x as its argument, has the dimension of the reciprocal of x. This can be seen from the normalization condition 4.17.

4.5.2 Properties of the Impulse Function

Some important properties of the impulse generalized function, that are useful to model linear systems, are listed in the following (Hsu, 1973a).

First, recall that $\delta(x) = u'(x)$. Then, from the properties of the unit step function

$$u(\alpha x) = \begin{cases} u(x), & \text{if } \alpha \text{ is positive,} \\ u(-x), & \text{if } \alpha \text{ is negative.} \end{cases} \quad (4.30)$$

Taking the derivative of the unit step function $u(\alpha t)$, one obtains

$$\frac{du(\alpha x)}{dx} = \begin{cases} \alpha\delta(x), & \text{if } \alpha \text{ is positive,} \\ -\alpha\delta(x), & \text{if } \alpha \text{ is negative.} \end{cases} \quad (4.31)$$

Therefore, combining both results, one obtains

$$\delta(\alpha x) = \frac{1}{|\alpha|}\delta(x). \quad (4.32)$$

Because it is possible to write $u(-t) = 1 - u(x)$, then

$$u'(-x) = -\delta(-x) = \frac{d[1 - u(x)]}{dx} = -\delta(x),$$

thus,

$$\delta(-x) = \delta(x). \quad (4.33)$$

By convention, one has the following property

$$x\delta(x) = 0. \quad (4.34)$$

The following property is a result of the sampling property, because the impulse is different from zero only at $x = 0$. It means that the area of the impulse takes the value of the function at $x = 0$.

$$f(x)\delta(x) = f(0)\delta(x), \quad (4.35)$$

if $f(x)$ is continuous at $t = 0$.

The last property can be generalized for any time $t = \tau$,

$$f(x)\delta(x - \tau) = f(\tau)\delta(x - \tau), \qquad (4.36)$$

and, for any continuous-time function $f(x)$,

$$f(x) = \int_{-\infty}^{\infty} f(\tau)\delta(x - \tau)\mathrm{d}\tau. \qquad (4.37)$$

The equation that defines the derivatives of the impulse function is

$$\int f(x)\delta^{(n)}(x)\mathrm{d}x = -\int f'(x)\delta^{(n-1)}(x)\mathrm{d}x. \qquad (4.38)$$

4.5.3 Composite Function with the Impulse

The composite function that results by combining the impulse with a function of time $f(x)$, that has N roots $x_i, i = 1, 2, \ldots, N$, is given by

$$\delta[f(x)] = \sum_{i=1}^{N} \frac{\delta(x - x_i)}{|f'(x_i)|}. \qquad (4.39)$$

This is an important result, that has several applications, mainly regarding the identification of zeros of a given function. It is demonstrated for a polynomial, but can be generalized for any function that has zeroes on the real line.

To demonstrate this last property of the impulse function for a polynomial, consider that the function $f(x)$ can be put in the following form, with no loss of generality,

$$f(x) = (x - x_1)(x - x_2) \cdots (x - x_N) = \prod_{i=1}^{N}(x - x_i).$$

in which t_i are the ordered ($t_1 < t_2 \cdots < t_N$) roots of the polynomial $f(x)$.

Figure 4.10 illustrates the composite function $\mathrm{u}[f(x)]$. Observe that this composite function is the sum of gate functions, of the form

$$\mathrm{u}[f(x)] = \sum_i g(x - x_i) = \sum_i [\mathrm{u}(x - x_i) - \mathrm{u}(x - x_{i+1})]. \qquad (4.40)$$

Note also, by the chain rule, that

$$\frac{\mathrm{d}}{\mathrm{d}x}\mathrm{u}[f(x)] = \delta[f(x)]f'(x). \qquad (4.41)$$

4.5 The Impulse Function

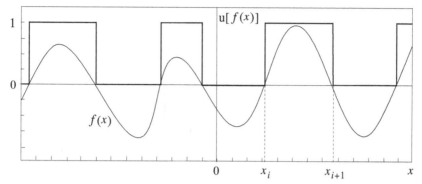

Figure 4.10 The function $u[f(x)]$.

Taking the derivative of the composite function in Equation (4.40) and equating to (4.41), yields

$$\delta[f(x)]f'(x) = \sum_i [\delta(x - x_i) - \delta(x - x_{i+1})], \quad (4.42)$$

which can be put into the following form

$$\delta[f(x)] = \sum_i \frac{[\delta(x - x_i) - \delta(x - x_{i+1})]}{f'(x_i)}. \quad (4.43)$$

This represents a series of alternating impulses that occur at the roots of the polynomial. Because the sign of each impulse coincides with the sign of the polynomial derivative at each point, it is possible to put the equation in its final form

$$\delta[f(x)] = \sum_i \frac{[\delta(x - x_i)]}{|f'(x_i)|}. \quad (4.44)$$

Example: Consider that $f(x) = x^2 - \alpha^2$, and determine the formula for $\delta[f(x)]$. This is an equation for a parabola that crosses the abscissa axis, as sketched in Figure 4.11.

Substituting the expression for $f(x)$ into the formula, one obtains

$$\delta[x^2 - \alpha^2] = \sum_{i=1}^{2} \frac{\delta(x - x_i)}{2|x_i|},$$

for $x_1 = -\alpha$ and $x_2 = \alpha$. Thus,

$$\delta[x^2 - \alpha^2] = \frac{1}{2|\alpha|} [\delta(x + \alpha) + \delta(x - \alpha)],$$

as can be observed in the figure. ∎

84 Generalized Functions

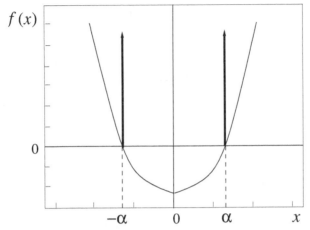

Figure 4.11 Parabola that crosses the abscissa at two points.

Example: Consider now that $f(x) = x^2 + 2x - 3$, and again determine the formula for $\delta[f(x)]$. For this polynomial, after determining the roots, one obtains $f(x) = (x-1)(x+3)$. Thus, $f'(x) = 2x + 2$, $f'(1) = 4$, and $f'(-3) = -4$.

Substituting the expression for $f(x)$, and the root values, into the formula, one obtains

$$\delta[x^2 + 2x - 3] = \sum_{i=1}^{2} \frac{\delta(x - x_i)}{4|x_i|},$$

therefore,

$$\delta[x^2 + 2x - 3] = \frac{1}{4}\left[\delta(x-1) + \delta(x+3)\right]. \blacksquare$$

Example: Determine an explicit formula for $\delta[f(x)]$, for the case $f(x) = \sin(x)$. Because the sine function has zeros at the points $x_k = k\pi$, one obtains

$$\delta[\sin(x)] = \sum_{k=-\infty}^{\infty} \delta(x - k\pi). \blacksquare$$

The mathematical theory of generalized functions has been extensively developed by mathematicians, engineers and theoretical physicists, in connection with the needs of theoretical and mathematical physics, engineering

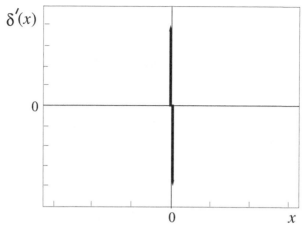

Figure 4.12 Doublet generalized function.

applications linked to Fourier and Laplace transforms, and the theory of integral and differential equations.

4.6 Doublet Generalized Function

The doublet generalized function $\delta'(x)$ is obtained as the derivative of the impulse function. It is shown in Figure 4.12.

$$\delta'(x) = \frac{\mathrm{d}\delta(x)}{\mathrm{d}x}. \tag{4.45}$$

The doublet function has the following fundamental property,

$$\int_{-\infty}^{\infty} f(x)\delta'(x)\mathrm{d}x = -\int_{-\infty}^{\infty} x'(x)\delta(x)\mathrm{d}x = -x'(0), \tag{4.46}$$

which can be obtained using integration by parts, for a function $f(x)$ which is continuous at $t = 0$ and vanishes outside some fixed interval.

Defining the n-th derivative of the impulse function as

$$\delta^{(n)}(x) = \frac{\mathrm{d}^n \delta(x)}{\mathrm{d}x^n}, \tag{4.47}$$

one obtains

$$\int_{-\infty}^{\infty} f(x)\delta^{(n)}(x)\mathrm{d}x = (-1)^n x^{(n)}(0). \tag{4.48}$$

86 *Generalized Functions*

Finally, observe that putting $f(x) = \tau\delta(\tau - x)$ and $n = 1$ into Equation (4.47), and integrating in τ, one obtains

$$\int_{-\infty}^{\infty} \tau\delta(\tau - x)\delta'(\tau)\mathrm{d}\tau = x\delta'(x), \tag{4.49}$$

thus,
$$x\delta'(x) = -\delta(x). \tag{4.50}$$

Also, putting $f(x) = x^n\delta(x)$ into Equation (4.46), yields

$$\int_{-\infty}^{\infty} x^n\delta(x)\delta'(x)\mathrm{d}x = -\int_{-\infty}^{\infty} nx^{n-1}\delta'(x)\delta(x)\mathrm{d}x. \tag{4.51}$$

Integrating again the second term $n - 1$ times, one obtains

$$\int_{-\infty}^{\infty} x^n\delta(x)\delta'(x)\mathrm{d}x = \int_{-\infty}^{\infty} (-1)^n n!\delta'(x)\delta(x)\mathrm{d}x, \tag{4.52}$$

therefore, observing the integrand, the following result is obtained

$$x^n\delta^{(n)}(x) = (-1)^n n!\delta(x). \tag{4.53}$$

Finally,
$$\delta'(-x) = -\delta'(x), \tag{4.54}$$

which means that the doublet generalized function is anti-symmetric.

4.7 The Ramp Function

The ramp function, $\mathrm{r}(x)$, is obtained as the integral of the unit step function, as sketched in Figure 4.13.

$$\mathrm{r}(x) = \int_{-\infty}^{x} \mathrm{u}(\tau)\mathrm{d}\tau. \tag{4.55}$$

It can also be written as $\mathrm{r}(x) = x\mathrm{u}(x)$.

Of course, by the Fundamental Theorem of the Calculus, the derivative of the ramp function gives the unit step function,

$$\frac{\mathrm{d}\,\mathrm{r}(x)}{\mathrm{d}x} = \mathrm{u}(x). \tag{4.56}$$

As can be seen, the ramp function grows without limit as time goes by. This function can be used to model different functions, including the

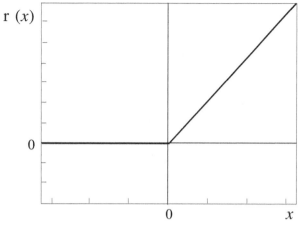

Figure 4.13 Ramp function.

sawtooth wave, that is used to synchronize the time base in oscilloscopes. The ramp function can also be used to produce composite functions, such as the half-wave signal that is produced as the result of the signal rectification by a semiconductor diode.

Example: Consider the function

$$f(x) = x^2 \operatorname{sgn}(x),$$

which is differentiable in the classical sense. Its derivative exists at all x, and is given by

$$f'(x) = 2|(x)| = 2[\operatorname{r}(x) + \operatorname{r}(-x)].$$

The second derivative, on the other hand, does not exist in the classical sense at $x = 0$, but the generalized derivative can be obtained, and gives

$$f''(x) = 2 \operatorname{sgn}(x). \blacksquare$$

4.8 The Exponential Function

The exponential function, depicted in Figure 4.14, is expressed as (Oppenheim et al., 2002)

$$f(x) = e^{\alpha x}, \tag{4.57}$$

in which α is a constant parameter. For $\alpha > 0$ the function grows without limit, and for $\alpha < 0$ it decreases to zero, and for $\alpha = 0$ the function is constant $f(x) = 1$, as times goes by.

88 Generalized Functions

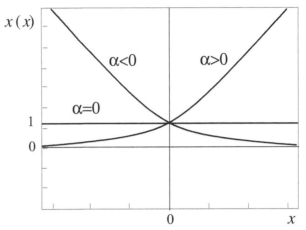

Figure 4.14 The exponential function.

This function has large application in Electrical Engineering, as well as, several other scientific areas. It is usually combined with the unit step function, to delimit its support. For instance, the decreasing exponential

$$y(x) = e^{-\alpha x} u(x), \tag{4.58}$$

for α positive, has its domain in the interval $[0, \infty]$.

Example: Compute the derivative of the following function, an exponential function limited to the interval $[0, \infty)$, using the Leibniz rule for the derivative of the product of functions. Gottfried Wilhelm Leibniz (1646–1716) was a German polymath, mathematician, philosopher, scientist, and diplomat, who developed the Calculus.

$$f(x) = e^{-\alpha x} u(x). \tag{4.59}$$

From the Leibniz rule of the derivative of the product of functions, it follows that

$$g(x) = f'(x) = [e^{-\alpha x}]' u(x) + e^{-\alpha x} u'(x). \tag{4.60}$$

Computing the indicated derivatives, one obtains

$$g(x) = -\alpha e^{-\alpha x} u(x) + e^{-\alpha x} \delta(x) = -\alpha e^{-\alpha x} u(x) + \delta(x), \tag{4.61}$$

which can be seen in Figure 4.15. Note that the abrupt transition of the exponential function at $x = 0$ produces an impulse at that point. ∎

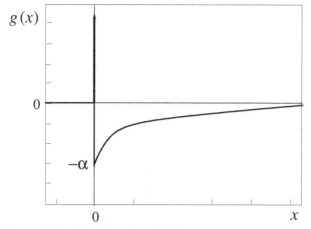
Figure 4.15 The derivative of the limited exponential function.

Example: Compute the nth derivative of the following function,

$$f(x) = e^{\alpha x} u(x). \tag{4.62}$$

Using the Leibniz rule of the derivative of the product of functions, and differentiating n times, one obtains the expression

$$f^{(n)}(x) = \alpha^n e^{\alpha x} u(x) + \sum_{k=1}^{n} \alpha^{k-1} \delta^{(k-1)}(x). \blacksquare \tag{4.63}$$

4.9 Discrete Functions

Several phenomena are modeled using discrete functions, for example the probability of calls placed at a telephone exchange, the number of objects produced by a factory per hour, the number of particles emitted by a radioactive sample, the number of people that enter a bank agency at a given date, or the output of a digital circuit.

Discrete time systems are characterized by difference equations, which express the relationship between discrete-time functions, or sequences. A sequence of numbers $\{x_k\}$ is an ordered collection, that is indexed by a set of integers $\{k\}$. Thus, if $k \in \mathbb{Z}$, then $x_k \in \{\ldots, x_{-2}, x_{-1}, x_0, x_1, x_2, \ldots\}$ (Gabel and Roberts, 1973).

4.9.1 Discrete Unit Step Function

For example, the discrete unit step function u(k) is the sequence $\{\ldots, 0, 0, 0, 1, 1, 1, \ldots\}$, in which the first one occurs at $k = 0$, as shown in Figure 4.16.

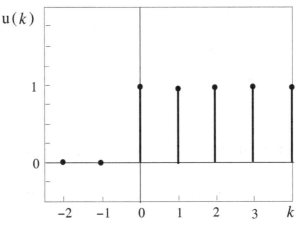

Figure 4.16 The discrete version of the unit step function.

4.9.2 Discrete Impulse Function

The discrete unit step function can be used to define the discrete impulse function, also known as Kronecker impulse, $\delta(k)$, after the German mathematician Leopold Kronecker (1823–1891), who made contributions to the fields of algebra and continuity of functions.

The discrete impulse function is defined as

$$\delta(k) = \begin{cases} 0 & \text{if } k \neq 0 \\ 1 & \text{if } k = 0, \end{cases} \tag{4.64}$$

For $k \in \mathbb{Z}$. Figure 4.17 sketches the discrete impulse function.

If one takes the difference between the discrete unit step function u(k) and a delayed version of it u($k-1$), the result is the Kronecker impulse function, as follows

$$\delta(k) = u(k) - u(k-1). \tag{4.65}$$

By the same token, the discrete unit step function can be obtained as

$$u(k) = \sum_{n=-\infty}^{k} \delta(n), \tag{4.66}$$

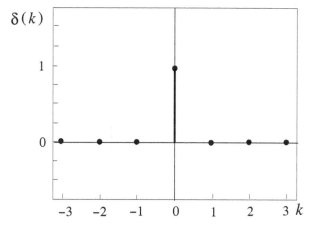

Figure 4.17 The discrete version of the impulse function.

in which the summation is done over the set of integers $\mathbb{Z}$.

4.9.3 Discrete Ramp Function

The discrete-time ramp, shown in Figure 4.18, can be written as the summation of the discrete unit step function, in the following manner

$$r(k+1) = \sum_{n=-\infty}^{k} u(n), \quad (4.67)$$

in which the summation is done over the set of integers $\mathbb{Z}$. The discrete ramp function can also be expressed as $r(k) = k u(k)$, for all $k \in \mathbb{Z}$.

On the other hand, the discrete unit step function can be obtained from the discrete ramp function, as indicated

$$u(k) = r(k+1) - u(k). \quad (4.68)$$

In fact, any arbitrary sequence can be represented as a linear combination of time-shifted Kronecker functions, as follows (Mandal and Asif, 2007)

$$f(k) = \sum_{n=-\infty}^{\infty} f(n)\delta(k-n), \quad (4.69)$$

in which, once more, the summation is done over the set of integers $\mathbb{Z}$. The summation 4.69 is usually taken as the starting point to define the discrete

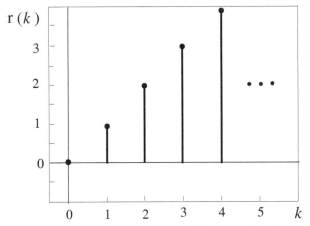

Figure 4.18   The discrete version of the ramp function.

convolution operation, which is one of the most important concepts in signal processing.

Example: The window function, $w(k)$, is used to filter a certain set of samples from a given function. It is defined as

$$w(k) = \begin{cases} 0 & \text{if } k < -n, k > n, \\ 1 & \text{if } -n \leq k \leq n. \end{cases} \quad (4.70)$$

For $k, n \in \mathbb{Z}$. Write this function using the discrete unit step function. The window function is used, for instance, to extract information from a given batch of data.

From the definition, as sketched in Figure 4.19 for $n = 2$, one can write the window function as

$$w(k) = u(k+n) - u(k-n). \blacksquare \quad (4.71)$$

Example: A decaying discrete-time exponential, defined for $k \in \mathbb{N}$, may be written as

$$f(k) = e^{-k} u(k). \quad (4.72)$$

The decaying exponential function is shown in Figure 4.20. This discrete function is largely used in signal processing and digital communications. $\blacksquare$

Example: A discrete-time complex exponential, which is shown in Figure 4.18, may be written as

$$y(k) = e^{j\omega k}, \quad (4.73)$$

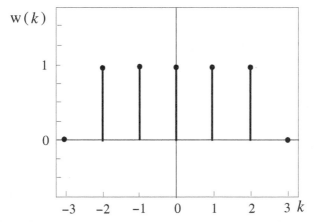

Figure 4.19 The discrete version of the window function, for $n = 2$.

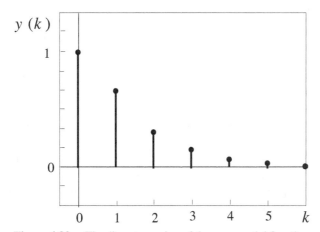

Figure 4.20 The discrete version of the exponential function.

and has the following property

$$y(k) = e^{j(\omega+2\pi)k} = e^{j\omega k}e^{j2\pi k} = e^{j\omega k}, \qquad (4.74)$$

because $e^{j2\pi k} = \cos(2\pi k) + j\sin(2\pi k) = 1$ for $k \in \mathbb{Z}$. ■

The discrete-time complex exponential represents a generalization of the sinusoidal function in the complex plane. As can be seen, the function is periodical and has modulus

$$|y(k)| = |e^{j\omega k}| = |\cos(2\pi k) + j\sin(2\pi k)| = 1, \text{ for } k \in \mathbb{Z}. \qquad (4.75)$$

94 *Generalized Functions*

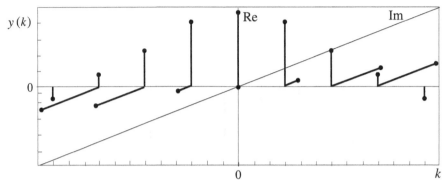

Figure 4.21 The discrete version of the complex exponential function.

4.10 Paul Dirac

Paul Adrien Maurice Dirac (1902–1984) is regarded as one of the most important physicists of the XX Century. He made fundamental contributions to the development of quantum mechanics and quantum electrodynamics, and formulated the equation which describes the behaviour of fermions and predicted the existence of antimatter. A picture of Dirac is shown in Figure 4.22.

Dirac, who graduated both in Electrical Engineering and Mathematics from the University of Bristol, shared the 1933 Nobel Prize in Physics with Erwin Schrödinger, and made significant contributions to harmonize the theories of general relativity and quantum mechanics.

Dirac also contributed a criticism about the political purpose of religion, at the 1927 Solvay Conference:

> "I cannot understand why we idle discussing religion. If we are honest – and scientists have to be – we must admit that religion is a jumble of false assertions, with no basis in reality. The very idea of God is a product of the human imagination. It is quite understandable why primitive people, who were so much more exposed to the overpowering forces of nature than we are today, should have personified these forces in fear and trembling. But nowadays, when we understand so many natural processes, we have no need for such solutions."

Figure 4.22 A portrait of Paul Adrien Maurice Dirac. Adapted from: Public Domain, http://nobelprize.org/ nobelprizes/physics/laureates/1933/dirac.html

4.11 Problems

1. Draw the graphs to obtain the following composite function $y(t) = u[\sin(\omega t)]$.

2. Obtain the graph for the following composite function $y(t) = \delta[\sin(\omega t)]$.

3. Prove that the impulse function can extract a sample of a given function, as in the following,

$$\int_{-\infty}^{\infty} f(x)\delta(x)\mathrm{d}x = f(0).$$

4. Demonstrate the equation that defines the derivatives of the impulse function,

$$\int f(x)\delta^{(n)}(x)\mathrm{d}x = -\int f'(x)\delta^{(n-1)}(x)\mathrm{d}x.$$

5. For the equation $f(x) = x^3 - x^2 - x + 1$, determine the formula for $\delta[f(x)]$ and sketch the graphs, considering that

$$\delta[f(x)] = \sum_i \frac{[\delta(x-x_i)]}{|f'(x_i)|}.$$

6. Compute the integral of the function $g(x) = [\delta(x+\alpha)+\delta(x-\alpha)]\mathrm{sgn}(x)$.

7. Using the Leibniz rule for the product of functions, determine the second derivative of the function

$$f(x) = e^{\alpha x} u(x).$$

8. Plot the following function, defined for $k \in \mathbb{N}$, obtain a formula for $f(k) = g(k) - g(k-1)$, and draw the graph.

$$g(k) = [1 - e^{-k}]u(k).$$

5
Probability Theory

"Probable impossibilities are to be preferred to improbable possibilities."
Aristotle

5.1 Reasoning in Games of Chance

The theory of probability evolved in a tortuous manner, from the first applications to games of chance to the beautiful mathematical developments based on set and measure theories. Probability theory began in France with studies about the chance in games. Antoine Gombaud (1607-1684), known as *Chevalier de Méré*, was a gentleman of considerable erudition, affectionate by card games, who used to discuss with Blaise Pascal (1623-1662) the probabilities of success in such games (Struik, 1987). Pascal, became interested in the subject and started a correspondence with Pierre de Fermat (1601-1665), em 1654, which gave birth to the very first ideas of finite probability (Zumpano and de Lima, 2004).

Girolamo Cardano (1501-1576) wrote the first known book on probability, entitled *De Ludo Aleae* (About Games of Chance). He was an Italian medical doctor and mathematician, and the book was published, in 1663, nine decades after his death. Cardano was a known card player, and the manuscript was a handbook on this subject, containing some discussion on probability.

Another Italian, the astronomer and physicist Galileu Galilei (1564-1642), was also interested in random events. Is a fragment of his work, written between 1613 and 1623, entitled *Sopra le Scorpete dei Dadi* (About the Games of Dice), Galileu answers a question posed by the Grand Duke of Tuscany: When three dice are thrown, even if the number nine, as well as, the number ten can be obtained in six different manners, in practice, the chance of obtaining nine is smaller than the chance of obtaining ten. How to explain that?

The first printed mathematical treatise on probability theory, published in 1657, was written by the Dutch scientist Christian Huygens (1629-1695). The manuscript, entitled *De Ratiociniis in Ludo Aleae* (About Reasoning in Games of Chance), was based on the concept of expectancy, or expected value, was published in 1657.

Abraham de Moivre (1667-1754) was an important mathematician who worked on the development of Probability Theory. He wrote a book of great influence in his time, called *Doctrine of Chances*. The Law of Large Numbers was discussed by Jacques Bernoulli (1654-1705) in his work Ars Conjectandi (The Art of Conjecturing).

Bernoulli, a Swiss mathematician of a family of prominent mathematicians, worked on the development of infinitesimal calculus, applying it to new problems. He published the first solution of a differential equation, that opened the path to the calculation of the variations of Leonhard Paul Euler (1707-1783) and Joseph Louis Lagrange (1736-1813), and extended their main applications to probability calculations (Dunham, 1990).

The study of probability was improved in the centuries XVIII and XIX, mainly because of the works of French mathematicians Pierre-Simon de Laplace (1749-1827) and Siméon Poisson (1781-1840), as well as the German mathematician Karl Friedrich Gauss (1777-1855) (de Laplace, 1814).

The work of Laplace defines a natural division in the history of probability, considering that his treaty Théorie Analytique des Probabilités, published in 1812, summed up most of the research on the subject of that time. Laplace also wrote a popular exposition to the educated public, entitled "A Philosophical Essay on Probabilities" (Parzen, 1979).

It worths mentioning the commentary by Poisson about the beginning of probability theory, in which he remarks a connection between Antoine Gombaud, the *Chevalier de Méré*, and Blaise Pascal (Struik, 1987),

"*Un problème relatif aux jeux de hasard, proposé à un austère janséniste par un homme du monde, a été l'origine du calcul de probabilités[1].*"

---

[1] A problem related to games of chance, proposed by a strict Jansenist (Pascal lived a simple and ascetic life in the Port Royal convent, since he was 25 years old) to a man of the world, was the origin of the calculus of probability

5.2 Measurable Space

A measurable space or probability space is a triple $(\Omega, \mathcal{F}, P)$, in which Ω is a non-empty set, called the universal set, or sample space, $\mathcal{F}$ is a σ-algebra, also called σ-field, a collection of subsets of Ω, containing the empty set $\emptyset$ and Ω itself, and closed under the formation of complements, countable unions and countable intersections. The measure P is defined in $\mathcal{F}$, as a measure from $\mathcal{F}$ to the set of real numbers.

If P is a probability measure, $(\Omega, \mathcal{F}, P)$ is called a probability space. For any set Ω, $\mathcal{F}$ is the family of all subsets of Ω. The countably additive probability measure P is a mapping from $\mathcal{F}$ to the set of real numbers $\mathbb{R}$, such that $0 \leq P(A) \leq 1$, for all $A \in \mathcal{F}$, with $P(\emptyset) = 0$, and $P(\Omega) = 1$ (Rosenthal, 2000).

A possible probability measure is defined in the following. Consider $\Omega = \{x_1, x_2, \dots\}$ a finite or countable infinite set, and the non-negative numbers $p_1, p_2, \dots$. Take $\mathcal{F}$ as the family of subsets of Ω and set

$$P(A) = \sum_{x_i \in A} p_i. \tag{5.1}$$

The set function P is a measure in $\mathcal{F}$ and one can define $P\{x_i\} = p_i$, $i = 1.2, \dots$. A probability measure is then obtained if

$$\sum_i p_i = 1. \tag{5.2}$$

That is, the set is normalized. If all $p_i = 1$, for instance, then P is called a counting measure.

Example: An engineer invites his fellow physicists and mathematicians to a party and, due to his popularity, he distributes invitations with identification badges. The badges bring numbers from the range $(0, 1)$, that is, decimals, with a special feature: mathematicians receive irrational numbers and physicists take the rational numbering. What is the probability of meeting a rational physicist at the party?

This problem is similar to the calculation of the Dirichlet function measure, discussed in the previous chapter, but with an associated probability measure. An alternative solution to the question considers that if each number is produced from a random experiment, such as to roll a die with ten faces, numbered from 0 to 9, an infinite number of times, then the occurrence of rational numbers is very unlikely.

Rational numbers always end in sequences of zeros or in repeating decimals, that is, always repeat a defined sequence after a certain number of digits. The probability of an infinite number of repetitions of the same sequence is practically zero. Therefore, the set of irrational mathematicians absorbs every measure of probability, and there is no chance of finding a rational physicist at the party.∎

5.2.1 Probability Measure

As discussed, a measure is a set function, that is, the assignment of a number $\mu(A)$ to each set A in a given class. A framework, as discussed in the previous chapters, must be imposed on the set class in which μ is defined, for example, the usual considerations of probability.

If Ω is the set whose points correspond to the possible outcomes of a random experiment, certain subsets of Ω are called events, which are assigned the probability measure. The attribution of the measure must follow a logical reasoning, as far as physical and intuitive as possible. If a result prompts the question, "$\omega \in A$?," then it is reasonable ask if "$\omega \in \overline{A}$?", among other questions.

This motivates the formation of a σ-algebra that is closed under the usual operations with sets, including the complement, the finite and enumerable union, and the finite and denumerable intersection. Since the answer to the question "$\omega \in \Omega$?" is always positive, therefore it is logical to conclude that Ω must also be an event of this algebra (Ash, 1972).

5.2.2 Probability Measure with the Riemann Integral

The Riemann integral, named after the German mathematician Georg Friedrich Bernhard Riemann (1826–1866), also has the necessary characteristics for a measure. Consider $(\mathbb{R}, \mathcal{B}(\mathbb{R}))$ the real line with its Borel set. Suppose there is a real function p on the line satisfying the following properties (Gray and Davisson, 1986):

$$p(x) \geq 0, \ x \in \Omega, \qquad (5.3)$$

$$\int_{\Omega} p(x) \mathrm{d}x = 1, \qquad (5.4)$$

this function has a well-defined integral on the set of reals and can be used in the definition of the probability measure.

For a set $A \in \mathcal{B}(\mathbb{R})$, in which $\mathcal{B}(\mathbb{R})$ represents the Borel family of line segments of the set of reals $\mathbb{R}$, one can define a measure P as the following

set function
$$P(A) = \int_A p(x)\mathrm{d}x. \tag{5.5}$$

Furthermore, if A_1 and A_2 are disjoint sets, by the linearity property,

$$P(A_1 \cup A_2) = \int_{A_1 \cup A_2} p(x)\mathrm{d}x = \int_{A_1} p(x)\mathrm{d}x + \int_{A_2} p(x)\mathrm{d}x = P(A_1) + P(A_2). \tag{5.6}$$

The definition of the measure can also be written using the indicator function $I_A(x)$, in the form

$$P(A) = \int p(x) I_A(x) \mathrm{d}x, \ A \in \mathcal{B}(R). \tag{5.7}$$

5.2.3 Probability Measure with the Lebesgue Integral

The measure developed by Henri Léon Lebesgue (1875-1941), discussed in a previous chapter, is a generalization of the concept of the length of an interval (a, b), which can be expressed as $\mu(a, b) = |b - a|$. A measure is always used in the context of an algebra, a family of sets with certain properties, and is typically positive or null, that is, $\mu \geq 0$. Furthermore, the measure is additive,

$$\mu\left(\bigcup_{i=1}^{\infty} F_i\right) = \sum_{i=1}^{\infty} \mu(F_i), \tag{5.8}$$

for a collection of measurable disjoint sets $\mathcal{F} = \{F_i\}$.

Examples of sets with Lebesgue measure zero, in $\mathbb{R}$, are the point and a set with a finite or countable number of points. Intervals usually have a non-zero measure in the set of reals.

Some properties of set functions, which are useful for defining probability measures, are demonstrated in the following. Let μ be a finitely additive set function, defined in the $\mathcal{F}$ field.

1. $\mu(\emptyset) = 0$.
2. $\mu(A \cup B) + \mu(A \cap B) = \mu(A) + \mu(B)$, for all $A, B \in \mathcal{F}$.
3. $\mu(A) = \mu(B) + \mu(A - B)$, if $A, B \in \mathcal{F}$ and $B \subset A$.
4. If μ is non-negative,
$$\mu\left(\bigcup_{i=1}^{n} A_i\right) \leq \sum_{i=1}^{n} \mu(A_i), \text{ for all } A_i \in \mathcal{F}.$$

5. If μ is a measure,
$$\mu\left(\bigcup_{i=1}^{\infty} A_i\right) \leq \sum_{i=1}^{\infty} \mu(A_i),$$
for all $A_i \in \mathcal{F}$ such that $\bigcup_{i=1}^{\infty} A_i \in \mathcal{F}$.

The properties are proven in the following.

1. Choose $A \in \mathcal{F}$ ouch that $\mu(A)$ is finite, then $\mu(A) = \mu(A \cup \emptyset) = \mu(A) + \mu(\emptyset)$, which results in $\mu(\emptyset) = 0$.

2. By the finite additivity property,
$$\mu(A) = \mu(A \cap B) + \mu(A - B),$$
$$\mu(B) = \mu(B \cap B) + \mu(B - A).$$
Adding the previous equations, one obtains
$$\mu(A) + \mu(B) = \mu(A \cap B) + [\mu(A - B) + \mu(B - A) + \mu(B \cap B)]$$
$$= \mu(A \cup B) + \mu(A \cap B).$$

3. It is possible to write $A = A \cup (A - B)$, therefore $\mu(A) = \mu(B) + \mu(A - B)$.

4. On obtains,
$$\bigcup_{i=1}^{n} A_i = A_1 \cup (\overline{A_1} \cap A_2) \cup (\overline{A_1} \cap \overline{A_2} \cap A_3) \cup \cdots \cup (\overline{A_1} \cap \cdots \cap \overline{A_{n-1}} \cap A_n).$$

The sets on the right-hand side of the equality sign are disjoint and
$$\mu(\overline{A_1} \cap \cdots \cap \overline{A_{n-1}} \cap A_n) \leq \mu(A_n).$$

Therefore,
$$\mu\left(\bigcup_{i=1}^{n} A_i\right) \leq \sum_{i=1}^{n} \mu(A_i).$$

5. For the case in which μ is a measure, it is possible to use the property

$$\bigcup_{n=1}^{\infty} A_i = \bigcup_{n=1}^{\infty} (\overline{A_1} \cap \cdots \cap \overline{A_{n-1}} \cap A_n).$$

In this manner, following the same line of thought,

$$\mu\left(\bigcup_{n=1}^{\infty} A_n\right) \leq \sum_{n=1}^{\infty} \mu(A_n).$$

A family of sets F_n is limited above by F, which is written as $F_n \uparrow F$, if $F_n \subset F_{n+1}$ and $\bigcup_{n \geq 1} F_n = F$. An basic property of a measure results from 5.8. If $F_1 \supset F_2 \supset \cdots$ is an increasing sequence of sets, that belong to the family $\mathcal{F}$, such that the union

$$F = \bigcup_{n=1}^{\infty} F_n$$

also belongs to $\mathcal{F}$, then

$$\mu(F) = \lim_{n \to \infty} \mu(F_n). \tag{5.9}$$

To demonstrate the property, it is sufficient to note that

$$F = F_1 \cup (F_2 - F_1) \cup (F_3 - F_2) \cup \cdots,$$

in which the sets $F_1, (F_2 - F_1), (F_3 - F_2), \cdots$ are disjoint, therefore

$$\mu(F) = \mu(F_1) + \mu(F_2 - F_1) + \mu(F_3 - F_2) + \cdots = \lim_{n \to \infty} \mu(F_n).$$

Some definitions are necessary to understand the theoretical concept of measurement. A set F is said not to be dense anywhere, if every neighborhood containing a point in F also has a point outside of F. A set is called closed if every neighborhood of a point x that contains a dot in F implies that $x \in F$. The integers on the real line are closed, that is, a point arbitrarily close to an integer is, in fact, an integer.

The Cantor set, after the Russian mathematician Georg Ferdinand Ludwig Philipp Cantor (1845-1918), shown in Figure 5.1, is closed, not dense anywhere and a subset of R, but it has non-zero measure. It is formed by iteratively partitioning a line segment $[0, 1]$, so that the central part is always removed from the remaining segments in the next step.

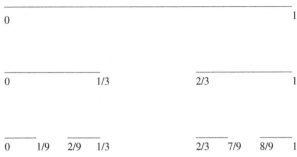

Figure 5.1 Measure of the Cantor set.

Since each x in the range $[0, 1]$ can be expressed as a decimal, of the type $0.x_1 x_2 x_3 \ldots$, the Cantor set is formed by all x whose ternary expansion does not contain the digit one, that is $x_i \neq 1$. As can be seen, the Cantor set contains an infinite number of straight segments, an impressive anticipation of the current fractal concept. However, its measure is null. Just notice that for each set $A_i \in \mathcal{F}$

$$\mu(A_1) = 1,$$
$$\mu(A_2) = 2/3,$$
$$\mu(A_3) = 4/9,$$
$$\vdots$$
$$\mu(A_n) = (2/3)^n,$$

therefore

$$\lim_{n \to \infty} \mu(A_n) = 0.$$

Despite the simplicity with which the Cantor set is defined, it is composed of an infinite family of line segments. And, moreover, it is not enumerable. Mathematical constructions like this have driven many mathematicians insane, including Cantor himself.

The Cantor set is certainly not empty, but nevertheless it has zero Lebesgue measure, in spite of being an uncountable set (Rosenthal, 2000). An explicit closed formula for the Cantor set is given by

$$C = \bigcap_{n=1}^{\infty} \bigcap_{k=0}^{3^{n-1}-1} \left(\left[0, \frac{3k+1}{3^n} \right] \cup \left[\frac{3k+2}{3^n}, 1 \right] \right). \tag{5.10}$$

From the notion of the Cantor set, it is possible to construct a really strange function, the Lebesgue's singular function, but better known as

"Devil's Staircase". It maps the domain $[0, 1]$ into the co-domain $[0, 1]$, it has the property of having an infinite number of steps between any steps, no matter how close they are, and the peculiarity of not being constant, but having null derivative at all points except in a set of zero measure (Bressoud, 2008).

The Lebesgue integral is defined for all Borel sets and it also becomes possible to interleave the order of limit and integration. For example, for a sequence of sets A_n that converges to A,

$$A = \lim_{n \to \infty} \bigcup_{i=1}^{n} A_n,$$

one has, for the measure P, defined previously,

$$P(A) = \int I_A(x) \mathrm{d}P(x) = \int \lim_{n \to \infty} I_{A_n}(x) \mathrm{d}P(x) = \lim_{n \to \infty} \int I_{A_n}(x) \mathrm{d}P(x)$$
$$= \lim_{n \to \infty} P(A_n),$$

and the set function is continuous.

5.3 The Axioms of Probability

The theory of probabilities is presented using three main approaches: the classical approach, the relative frequency approach and the axiomatic approach. The classical approach is based on the symmetry of the experiment. This causes a problem, because the concept of probability is used to elaborate the mental experiment, in a cyclical way, because it is only defined for equiprobable events (Alencar, 2008e).

The relative frequency approach, developed by Richard Edler von Mises (1883-1953) at the beginning of the XX Century, is more recent and can be justified with the realization of a grand experiment to test the convergence of the series of events. But, because the events are random, it is not possible to prove the convergence, unless with the realization of an infinite number of experiments.

Considering the mentioned problems with both approaches, only the axiomatic approach is presented in this book. The readers who are interested in the classical or relative frequency approaches are invited to review the commented bibliography in Appendix C.

The basic axioms of probability were established by Andrei Nikolaevich Kolmogorov (1903-1987), and allowed the development of the complete theory. Kolmogorov's first paper on probability appeared, in 1925, it was published jointly with Aleksandr Yakovlevich Khinchin (1894-1959), and contains the results on inequalities of partial sums of random variables which became the basis for martingale inequalities and the stochastic calculus.

Khinchin published several papers on measure theory and, in 1927, collected his contributions to this area in the paper *Recherches sur la Structure des Fonctions Mesurables*, published in the journal *Fundamenta Mathematicae*, of the Polish Academy of Sciences. This paper came out the same year he was appointed as a professor at Moscow University. In the same year, he also developed the basic laws of probability theory. In 1934, Khinchin published the foundations for the theory of stationary random processes in the journal *Mathematische Annalen*.

Kolmogorov was appointed a professor at Moscow University, in 1931. His thesis on probability theory, better known in its German translation from Russian, *Grundbegriffe der Wahrscheinlichkeitsrechnung* (Foundations of the Theory of Probability), published in 1933, put probability theory in a rigorous way, based on fundamental axioms.

In this work Kolmogorov managed to combine Advanced Set Theory, developed by Cantor, with Measure Theory, established by Lebesgue, to produce what is known as the modern approach to Probability Theory. The three fundamental statements are as follows (Papoulis, 1983):

- Axiom 1 – $P(\Omega) = 1$, in which Ω denotes the sample space or universal set and $P(\cdot)$ denotes the associated probability;
- Axiom 2 – $P(A) \geq 0$, in which A denotes an event belonging to the sample space;
- Axiom 3 – $P(A \cup B) = P(A) + P(B)$, in which A and B are mutually exclusive events and $A \cup B$ denotes the union of events A and B.

The third axiom is usually split into two more specific ones:

- Axiom 3a – The measure of probability of the finite union of sets is given by

$$P\left(\bigcup_i^N A_i\right) = \sum P(A_i), \qquad (5.11)$$

taking into account that $A_i \in \mathcal{F}$ are mutually exclusive events, that is, $A_i \cap A_j = \emptyset$ for $i \neq j$.

- Axiom 3b – For events obtained from an infinite family of sets, that is, from a σ-algebra, it is necessary to consider the denumerable union

$$P\left(\bigcup_i^\infty A_i\right) = \sum P(A_i), \quad (5.12)$$

in which, again, A_i are mutually exclusive events.

The first three axioms form, from the algebra of sets or events, a probability space, which is a measurable space with a measure P that obeys the rule $0 \leq P(A) \leq 1$. The probability measure is always applied to sets.

The probabilistic space is characterized by a triple $(\Omega, \mathcal{F}, \mathcal{P})$, in which Ω is the universal set, $\mathcal{F}$ is a family of subsets of the universal set of interest for the theoretical experiment, and P is the probability measure, that can be applied to the elements of $\mathcal{F}$.

The inclusion of the last axiom originates a σ-algebra, that is a mathematical structure that is capable of dealing with the denumerable infinite union of sets, that is, with the notion of limit of a sequence of sets.

Kolmogorov established a firm mathematical basis, with the axiomatic approach, on which other theories rely, including the Theory of Stochastic Processes, the Communications Theory, the Complexity Theory and the Information Theory, that use his axiomatic approach to Probability Theory.

His fundamental work in this area was published in 1933, in Russian, and soon afterwards, in German, *Grundbegriffe der Wahrscheinlichkeits Rechnung* (Foundations of the Theory of Probability) James (1981). He unified the Advanced Theory of Sets, developed by Georg Cantor, with the Measure Theory, developed by Henri Lebesgue, to produce the modern axiomatic approach of the Probability Theory.

The application of the axioms makes it possible to deduce all results relative to Probability Theory. For example, the probability of the empty set, $\emptyset = \{\}$, can be calculated as follows. First, notice that

$$\emptyset \cup \Omega = \Omega,$$

since the sets $\emptyset$ and Ω are disjoint. Thus it follows that

$$P(\emptyset \cup \Omega) = P(\Omega) = P(\emptyset) + P(\Omega) = 1 \Rightarrow P(\emptyset) = 0.$$

In the case of sets A and B which are not disjoint it follows that

$$P(A \cup B) = P(A) + P(B) - P(A \cap B).$$

5.4 Axioms of the Expectation Operator

There is an alternative approach, that is not followed in this text, to present the theory of probability in which the theory is based on the axiomatization of the concept of expectation, rather than of that of a probability measure. The main reasons that justify that approach are related to familiarity of some people with the concept of average value, which allows the presentation of topics such as optimization and approximation, among others (Whittle, 1970).

The expectation operator, $\mathbb{E}[X]$. can be defined as the mean value of a random variable $X(\omega)$, which is a function of ω, an outcome of the universal set Ω. The expectation is the ideal average value associated to the random variable, taken over all the outcomes of a theoretical experiment. It is expected that the sample averages converge to the expected value, to justify and give self-consistency to that approach. The expectation operator obeys the following set of rules:

- Axiom 1 – If $X \geq 0$ then $\mathbb{E}[X] \geq 0$.
- Axiom 2 – $\mathbb{E}[\alpha X] = \alpha \mathbb{E}[X]$, for a constant α.
- Axiom 3 – $\mathbb{E}[X + Y] = \mathbb{E}[X] + \mathbb{E}[Y]$, for any random variables X and Y.
- Axiom 4 – $\mathbb{E}[1] = 1$.
- Axiom 5 – If a sequence of random variables $\{X_n(\omega)\}$ increases monotonically to a limit $X(\omega)$, then $\mathbb{E}[X] = \lim \mathbb{E}[X_n]$.

Therefore, the expectation is a positive, linear operator, with the normalization $\mathbb{E}[1] = 1$, for which the average and limiting operations commute.

Example: Suppose that the sample space is the real line, which means that ω is a real number. Then, it is possible to write

$$\mathbb{E}[X] = \int_{-\infty}^{\infty} X(\omega) p(\omega) d\omega.$$

Consider that the measure p obeys the conditions:

$$\int_{\Omega} p(\omega) d\omega = 1,$$

and

$$p(\omega) \geq 0.$$

Therefore, it is possible to obtain a formula for the probability of the event A, as follows

$$P(A) = \mathbb{E}[I_A] = \int_A p(\omega)\mathrm{d}\omega,$$

in which I_A is the indicator function of the set A. In this case, $p(\omega)$ is regarded as a probability density function on Ω, or a continuous probability distribution on the sample space Ω.

5.5 Bayes' Theorem

Thomas Bayes (1701-1761) was an English statistician, philosopher and Presbyterian minister, who formulated a special case of the theorem that bears his name. Bayes never published the theorem, but his notes were edited and published, in "An Essay Towards Solving a Problem in the Doctrine of Chances", which was read to the Royal Society of London, in 1763, after his death, by Richard Price (1723-1791), a Welsh moral philosopher and nonconformist preacher (Bayes, 1763).

Bayes' theorem was further improved by Pierre-Simon, marquis de Laplace (1749-1827), and defines the probability of an event, based on conditions that are given about that event. Bayes' rule, which is essential for the development of Information Theory, is a method to calculate conditional probabilities.

To compute the conditional probability, the sample space is reduced to the set B. Then, the probability associated to the joint occurrence of A and B, divided by the probability of event B, gives the conditional probability of A given B, as illustrated in Figure 5.2.

$$P(A|B) = \frac{P(A \cap B)}{P(B)}. \qquad (5.13)$$

assuming $P(B) \neq 0$.

An equivalent manner of stating the same result is the following,

$$P(A \cap B) = P(A|B) \cdot P(B), \quad P(B) \neq 0. \qquad (5.14)$$

Some important properties of sets are presented next, in which A and B denote events from a given sample space.

- If A is independent of B, then $P(A|B) = P(A)$. It then follows that $P(B|A) = P(B)$ and that B is independent of A.

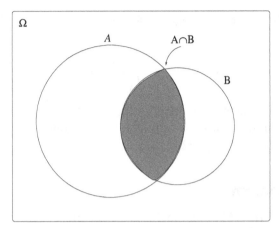

Figure 5.2 Venn diagram of the intersection of sets A and B.

- If $B \subset A$, then: $P(A|B) = 1$.
- If $A \subset B$, then: $P(A|B) = \frac{P(A)}{P(B)} \geq P(A)$.
- If A and B are independent events then $P(A \cap B) = P(A) \cdot P(B)$.
- If $P(A) = 0$ or $P(A) = 1$, then event A is independent of itself.
- If $P(B) = 0$, then $P(A|B)$ can assume any arbitrary value. Usually in this case one assumes $P(A|B) = P(A)$.
- If events A and B are disjoint, and non-empty, then they are dependent.

A partition is a possible splitting of the sample space into a family of subsets, in a manner that the subsets in this family are disjoint and their union defines the sample space. It follows that any set in the sample space can be expressed with the use of a partition of that sample space and, therefore, be written as a union of disjoint events.

The following property can be illustrated by means of the Venn diagram, as illustrated in Figure 5.3.

$$B = B \cap \Omega = B \cap \cup_{i=1}^{M} A_i = \cup_{i=1}^{N} B \cap A_i.$$

It now follows that

$$P(B) = P(\cup_{i=1}^{N} B \cap A_i) = \sum_{i=1}^{N} P(B \cap A_i),$$

$$P(A_i|B) = \frac{P(A_i \cap B)}{P(B)} = \frac{P(B|A_i) \cdot P(A_i)}{\sum_{i=1}^{N} P(B \cap A_i)} = \frac{P(B|A_i) \cdot P(A_i)}{\sum_{i=1}^{N} P(B|A_i) \cdot P(A_i)}.$$

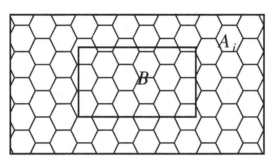

Figure 5.3 Partition of set B by a family of sets $\{A_i\}$.

Bayes' theorem is useful in several areas of study. It can be used to test hypotheses, such as those established for the occupancy of blank spaces in cognitive wireless sensor networks (Braga et al., 2015), or to evaluate the likelihood of certain diseases, for example (Joyce, 2008).

Example: If B is an arbitrary event, then, $P(\emptyset|B) = 0$ and $P(\Omega|B) = 1$, that is, the probabilities of the impossible event and of the sure event are not affected by the occurrence of B.

Aside from several other applications, the conditional probability forms the basis to understand information transmission by a noisy channel, in which A is associated to the source information and B if associated to the received information, after it passes through the channel and suffers the influence of the additive noise. ∎

Example: If B is an event with zero probability, and A is an arbitrary event, then $P(A \cap B) = 0 = P(A) \cdot P(B)$, that is, A and B are independent events. Events with zero probability are, therefore, independent of any other events. In fact, they are independent events, and Ω is independent of all other events. ∎

Any event can be mounted on a partition set and, in this manner it becomes a family of disjoint sets, that is, formed by events that are effectively disjoints, as shown in Figure 5.4.

The set B can be written as its own intersection with the universal set. Thus, considering that the latter can be divided into a partition of disjoint sets,

$$B = B \cap \Omega = B \cap \cup_{i=1}^{N} A_i = \cup_{i=1}^{N} B \cap A_i.$$

112 Probability Theory

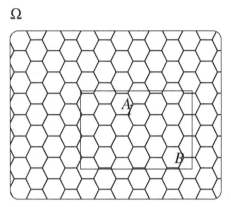

Figure 5.4 Partition of a set.

Therefore, the probability of occurrence of event B can be written as the sum of the probabilities associated to the disjoint sets

$$P(B) = P(\cup_{i=1}^{N} B \cap A_i) = \sum_{i=1}^{N} P(B \cap A_i). \qquad (5.15)$$

The joint probability of A_i given the occurrence of B, can be put in the form

$$P(A_i|B) = \frac{P(A_i \cap B)}{P(B)} = \frac{P(B|A_i) \cdot P(A_i)}{\sum_{i=1}^{N} P(B \cap A_i)}.$$

Using the previous result, on obtains

$$P(A_i|B) = \frac{P(B|A_i) \cdot P(A_i)}{\sum_{i=1}^{N} P(B|A_i) \cdot P(A_i)}. \qquad (5.16)$$

For any countable sequence of disjoint sets , A_n, one has

$$\begin{aligned}
P\left(\bigcup_n A_n \Big| B\right) &= \frac{1}{P(B)} \cdot P\left(\left(\bigcup_n A_n\right) \cap B\right) \\
&= \frac{1}{P(B)} \cdot P\left(\bigcup_n A_n \cap B\right) \\
&= \frac{1}{P(B)} \cdot \left(\sum_n P(A_n \cap B)\right) \\
&= \sum_n P(A_n|B). \qquad (5.17)
\end{aligned}$$

Finally, if A and B are independent events, the the events A and $\overline{B}$ are also independent, as the events $\overline{A}$ and B, and the events $\overline{A}$ and $\overline{B}$. In fact, all the sets from $\{\emptyset, A, \overline{A}, \Omega\}$ are independent of all the sets in $\{\emptyset, B, \overline{B}, \Omega\}$.

Example: Verify the independence between A and $\overline{B}$, given that A ans B are independent.

$$\begin{aligned} P\left(A \cap \overline{B}\right) &= P\left(A \cap (\Omega - B)\right) = P(A) - P(A \cap B) \\ &= P(A) - P(A) \cdot P(B) \\ &= P(A)(1 - P(B)) = P(A) P\left(\overline{B}\right). \blacksquare \end{aligned} \qquad (5.18)$$

5.6 Andrei Kolmogorov

Andrei Nikolaevich Kolmogorov, whose picture is shown in Figure 5.5, was born in Tambov, Russia on April 25 1903, and died in Moscow on October 20, 1987. In his fruitful life, Kolmogorov published over 300 scientific articles and reports and formed more than 60 doctoral researchers.

The list of Kolmogorov's contributions to science is indeed immense, and goes from trigonometric series to logic, passing through integration theory, topology, automata theory, set theory, measure theory, dynamic systems, statistics, stochastic processes, theory of algorithms, linguistics, turbulence

Figure 5.5 Andrei Kolmogorov was one of the more important scientists of the 20th Century (Source: Konrad Jacobs/Wikimedia Commons).

theory, celestial mechanics, equations differentials, ballistics, geology, metal crystallization, just to name a few areas.

However, he became best known for his participation in the development of two important fields of research: the Axiomatic Theory of Probability and the Complexity Theory.

The relative frequency approach is more recent. It has been developed by Richard Edler von Mises in the early 20th Century, and justifies the probability by the exhaustive repetition of the random experiment to obtain statistical regularity. Probability, according to this theory, is the limit of the relative frequency of successes obtained when performing a certain random experiment (Alencar, 2008e).

Kolmogorov Complexity is an algorithmic theory of information, that deals with the amount of information of mathematical entities, measured by the dimension of its smallest algorithmic description. It is a computation version of Information Theory, developed by Claude E. Shannon in the 1940s.

It was developed by Kolmogorov in the 1960s, but also has contributions from Ray J. Solomono and Gregory Chaitin, and represents a concept of randomness for sequences. The Turing Machine, the simplest conceptual computer, is used to describe the sequences. This machine was proposed, in 1936, by Alan Turing to mathematically formalize the concept computer, and contains a tape on which a head can write symbols of a certain set. The calculations are made with the movements of the head over the tape.

Kolmogorov Complexity quantifies information based on the length of a program that calculates a given sequence in the Turing machine, using binary digits for its representation. It is also possible to define conditional complexity, in a similar way to information, or conditional entropy of Shannon's theory, such as the length of the program that calculates information from a given program entry.

A curious aspect of Andrei Kolmogorov's life was his dedication to teach Mathematics and other subjects for the fundamental level. He even published a book on the subject and also used to teach children for many years (Alencar, 2008f).

5.7 Problems

1. Obtain $P(A)$, using the following definition of measure, for the set $A = (-\infty, x]$,
$$P(A) = \int p(x) I_A(x) \mathrm{d}x, \ A \in \mathcal{B}(R).$$

2. Prove that if A and B are independent events then $P(A \cap B) = P(A) \cdot P(B)$.

3. demonstrate that if events A and B are disjoint, and non-empty, then they are dependent.

4. Prove that if B is an arbitrary event, then, $P(\emptyset|B) = 0$ and $P(\Omega|B) = 1$, that is, the probabilities of the impossible event and of the sure event are not affected by the occurrence of B.

5. If A and B are independent events, verify that all the sets from $\{\emptyset, A, \overline{A}, \Omega\}$ are independent of all the sets in $\{\emptyset, B, \overline{B}, \Omega\}$.

6

Random Variables

"Nature laughs at the difficulties of integration."
Pierre-Simon de Laplace

6.1 The Concept of a Random Variable

The idea of a random variable has the objective of transposing events of the sample space Ω to Borel segments, or rectangles, in the set of real numbers. This concept is defined informally, by many authors, so that readers can become acquainted with the idea before the appearance of formal applications in the theory.

In reality, what is called a random variable is not a variable, in the strict sense of the word, because it represents a function of random events, that are defined as a result of operations carried out in an algebra of sets. It could be more adequately defined as a dependent variable, the term independent variable is considered as representing the result of the experiment.

In addition, the variable is also not random, because it is a well-defined function, that is deterministic in this domain, and it usually has an inverse. However, the term random variable, although inaccurate, is still used in the literature.

The mapping $X : \Omega \mapsto \mathbb{R}$ is completely deterministic. This way, $X = f(\omega)$, in which $\omega \in \Omega$, as shown in Figure 6.1. All the events of the sample space now have a counterpart in the set of real numbers, which is more adequate for usual operations, such as limit and integration.

In the case of probability spaces, the expression "random variable" means a measurable function. That is, if $(\Omega, \mathcal{F}, P)$ is a probability space, then $X : \Omega \mapsto \mathbb{R}$ is a random variable if, for every $x \in \mathbb{R}$, the set $X^{-1}([x, \infty])$ is in $\mathcal{F}$, that is, (Rosenthal, 2000)

$$\{\omega \in \Omega : X(\omega) \geq x\} \in \mathcal{F}. \tag{6.1}$$

118   Random Variables

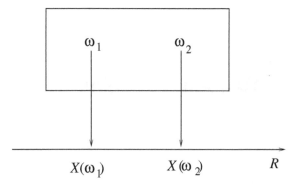

Figure 6.1   Mapping of the events in the random variable domain.

In the case in which $\Omega \subset \mathbb{R}$ is a measurable set, and $\mathcal{F} = \mathcal{B}$ is the σ-algebra of Borel subsets of Ω, the random variables are exactly the Borel functions $\mathbb{R} \mapsto \mathbb{R}$ (Capiński and Kopp, 2005).

Example: The experiment of rolling a dice and observing the result can be mapped into the set of natural numbers, $\mathbb{N}$, in the following manner

$$X(\omega_1) \to 1, X(\omega_2) \to 2, X(\omega_3) \to 3, X(\omega_4) \to 4, X(\omega_5) \to 5, X(\omega_6) \to 6.$$

because the function $X(\cdot)$ maps each outcome into a symbol of the set of natural numbers. ∎

This allows for mathematical operations and also makes it possible to plot graphs, and therefore results can be analyzed. When analyzing the problem, it is important to distinguish between the elements of the sample space $\Omega = \{\text{heads}, \text{tails}\}$, which only indicate the sides of the coin and, therefore, it does not make sense to assign probabilities to them, and the elements of the algebra, or family of subsets in the sample space, $\mathcal{F} = \{\emptyset, \Omega, \{\text{heads}\}, \{\text{tails}\}\}$, that represent the result of the experiment "tossing a coin and writing down the exposed side", to which is assigned the probability measure P. For example, $P(\{\text{heads}\}) = P\{\text{tails}\} = 1/2$.

Following what has been discussed, if f is a Borel function, defined in the considered space, and X is a random variable, then the function $f(X)$ is also a random variable (Rényi, 2007). To prove the statement, is suffices to consider, using a property of inverse functions, that $f^{-1}(X)(\mathcal{B}(\mathbb{R})) = X^{-1}\left[f^{-1}(X)(\mathcal{B}(\mathbb{R}))\right] \subset X^{-1}(\mathcal{B}(\mathbb{R})) \in \mathcal{F}$. Therefore, any function of a random variable is also a random variable.

6.1.1 Algebra Generated by a Random Variable

If X is a random variable, then $X^{-1}(\mathcal{B}) \subset \mathcal{F}$, but it can be a subset with a much smaller cardinality, depending on the complexity of X. This σ-algebra can be denoted as $\mathcal{X}$ and can be called the σ-algebra generated by X. The simplest case is that in which X is a constant, $X = c$, that produces a so called degenerate probability distribution. So, for a given set $B \in \mathcal{B}$, then $X^{-1}(\mathcal{B})$ is Ω or $\emptyset$, depending if $c \in B$ or not, and the σ-algebra generated is the trivial, $\mathcal{F} = \{\emptyset, \Omega\}$.

If X assumes different two values, a and b, then $\mathcal{X}$ contains four elements, $\mathcal{X} = \{\emptyset, \Omega, X^{-1}(\{a\}), X^{-1}(\{b\})\}$. If X can be one of many finite values, then $\mathcal{X}$ is finite. If X assumes countable values, then $\mathcal{X}$ is non-countable and can be identified with the σ-algebra of all the subsets of a countable set. One may notice that the cardinality of $\mathcal{X}$ increases with the complexity of X.

It must be considered that the set of points in which leaps occur in a probability distribution is countable, which allows the attribution of a probability to all the points considered, and that the distribution function that increases only through leaps is called a discrete distribution. In addition, the distributions always have limits to the left, even if they do not coincide with the limits to the right of the discontinuity points.

Example: Consider $X(\omega) = \omega$ and $Y(\omega) = 1 - \omega$, for $\Omega = [0, 1]$. Then, $\mathcal{X} = \mathcal{Y} = \mathcal{F}$, and a σ-algebra cannot be independent from itself, apart from the trivial case.

Let $A \in \mathcal{F}$, then the criterion of independence requires $P(A \cap A) = P(A) \times P(A)$, that is, the set A belongs to both σ-algebras. But this implies $P(A) = P(A)^2$, which can only happen if $P(A) = 1$ or $P(A) = 0$. Therefore, a σ-algebra independent from itself consists of sets with measure one or zero. ∎

It is important to mention that, although the distribution is the most important concept to characterize a random variable, it does not specify it completely. If, for example, the distributions of the random variables X and Y are known, it is not possible, in general, to determine the distribution of the variable $Z = f(X, Y)$ only from this information. For this, it is necessary to known the joint distribution of X and Y (Rényi, 2007).

6.1.2 Lebesgue Measure and Probability

The simple Lebesgue measure, previously discussed, can be used in the calculation of probabilities in the case of uniform distribution in the interval $U = [0, 1]$, defined in $\mathbb{R}$, because in this case the probability measure is given by the length of the variable's interval exactly (Rosenthal, 2000). That is, $P(0 \leq X \leq 1/2) = 1/2$ and, in general for the uniform distribution, $P(a \leq X \leq b) = |b-a|, 0 \leq a \leq b \leq 1$.

Apart from this, for the disjoint subsets of U, $A = [a, b]$ and $B = [c, d]$, in which $0 \leq a \leq b \leq c \leq d \leq 1$, it is always true that $P(A \cup B) = |b - a| + |d - c|$, that is, finite additivity can be applied,

$$P(A \cup B) = P(A) + P(B). \qquad (6.2)$$

Evidently, to allow for countable operations with sets, such as limits, which are important in probability theory, it is necessary to extend Equation (6.2) to a countably infinite number of disjoint sets,

$$P(A_1 \cup A_2 \cup A_3 \cup \cdots) = P(A_1) + P(A_2) + P(A_3) + \cdots, \qquad (6.3)$$

which is called countable additivity. Uncountable additivity is not defined, because it would imply, for example,

$$P([0,1]) = \sum_{x \in [0,1]} P(\{x\}), \qquad (6.4)$$

which is clearly false, because the left-hand side of the equation is equal to 1, while the right-hand side is 0, considering that the measure of a point is zero (Rosenthal, 2000).

The following sections present a probability measure valid for any distribution, as long as the variable is defined in a Borel interval, or rectangle, in the set of real numbers. It is initially defined for an open subset of $\mathbb{R}$.

6.2 Cumulative Distribution Function

The Cumulative Distribution Function (CDF), represented in Figure 6.2, is an application of the probability measure, in the probability space $(\mathbb{R}, \mathcal{B}(\mathbb{R}), P)$, to a specific set, that is,

$$P_X(x) = P\{x \in \mathbb{R} : -\infty < X \leq x\} = P\{X \leq x\}. \qquad (6.5)$$

Note that this definition is equivalent to that of a measurable function, induced by the measure in the probability space $(\Omega, \mathcal{F}, P)$, in which the reuse of P denotes a certain overindulgence of the language, and the simplification

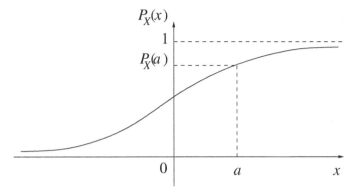

Figure 6.2 Cumulative distribution function.

$P(\{X \leq x\}) = P\{X \leq x\}$ is made (Marques, 2009),

$$P_X(x) = P\{\omega \in \Omega : -\infty < X(\omega) \leq x\}$$
$$= P\{X^{-1}((-\infty, x]) \in \mathcal{F}, \forall x \in \mathbb{R}\}. \quad (6.6)$$

With the definition, a function in the set of real numbers can be obtained, instead of a mathematical operation between generic functions, which makes the process of measuring probability more operational and appropriate for calculating the means, among other operations.

The CDF has the following properties, which can be verified:

1. All the probabilities are exhausted in the universal set

$$P_X(\infty) = P\{-\infty < X \leq \infty\} = P\{X < \infty\} = P(\Omega) = 1; \quad (6.7)$$

2. The probability of a point is zero,

$$P_X(-\infty) = P\{-\infty < X \leq -\infty\} = P\{X < -\infty\} = P(\emptyset) = 0; \quad (6.8)$$

3. For the disjoint Borel segments, like

$$\{X > x\} \cup \{X \leq x\} = \Omega,$$

that form a partition, then

$$P\{X > x\} + P\{X \leq x\} = 1 \implies P\{X > x\} = 1 - P_X(x); \quad (6.9)$$

122 Random Variables

4. The probability that a random variable is in the interval $[a,b]$ is given by

$$P\{a < X \leq b\} = P\{X \leq b\} - P\{X \leq a\} = P_X(b) - P_X(a). \quad (6.10)$$

The CDF $P_X(x)$ can be written as the primitive of a function, that is

$$P_X(x) = \int_{-\infty}^{x} p_X(\alpha) d\alpha.$$

Therefore, $P_X(a)$ represents the area under the curve up to point a. From the fundamental theorem of Calculus, it can be verified that

$$p_X(x) = \frac{d}{dx} P_X(x),$$

in which $p_X(x)$ is known as the probability density function (pdf), illustrated in Figure 6.3. In this figure, the darkened area represents the calculation of the probability density function.

It is important to notice that the CDF provides a measure of probability; therefore, it is dimensionless. However, the pdf does have dimension, which is the inverse of the dimension of the random variable considered.

Example: The exponential probability density models packet arrival time in a computer network and, also, the duration of a call in a telephone system. This distribution is generally written as

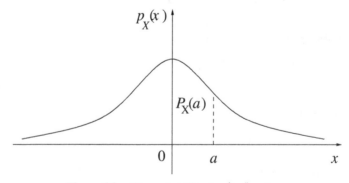

Figure 6.3   The probability density function.

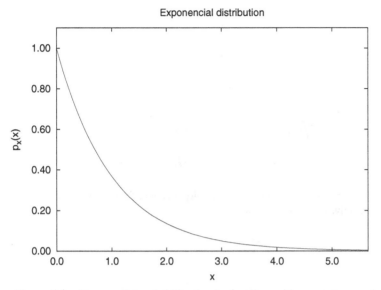

Figure 6.4 Exponential probability density function, with parameter $a = 1$.

$$p_X(x) = \alpha e^{-\alpha x} u(x),$$

in which u(x) is the unitary step function, shown in Figure 6.4.

An exponential random variable X, with parameter α, has $P\{X \geq x\} = e^{-\alpha x}$, which means that the probability that the variable exceeds a given value x is also exponential. This distribution has the property of being memoryless, that is, $P\{X > t + s | X > t\} = P\{X > s\}$. ∎

Next, some important relations between the CDF and the pdf are presented:

$$P\{X > x\} = 1 - \int_{-\infty}^{x} p_X(\alpha) \, d\alpha = \int_{-\infty}^{\infty} p_X(\alpha) \, d\alpha - \int_{-\infty}^{x} p_X(\alpha) \, d\alpha$$
$$= \int_{x}^{\infty} p_X(\alpha) \, d\alpha$$

$$P\{a < X \leq b\} = P_X(b) - P_X(a) = \int_{-\infty}^{b} p_X(x) \, dx - \int_{-\infty}^{a} p_X(x) \, dx$$
$$= \int_{a}^{b} p_X(x) \, dx$$

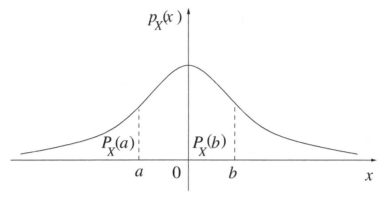

Figure 6.5 The area under the probability density function, from a to b is the difference between the respective cumulative distributions.

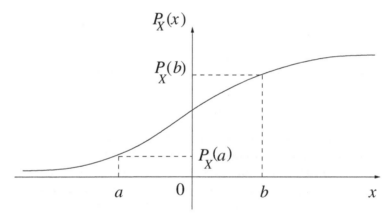

Figure 6.6 The probability measurement using the cumulative distribution function.

Example: A voice signal can be modeled using a Laplace distribution, as follows, for which the previous property can be verified

$$p_V(v) = \frac{a}{2}e^{-a|v|}. \blacksquare$$

The Laplace distribution was named after the French mathematician, astronomer and physicist Pierre-Simon, Marquis de Laplace (1749-1827),

In the generic case, the Laplace distribution has two control parameters, to adjust the mean and the standard deviation of the distribution.

6.2 Cumulative Distribution Function

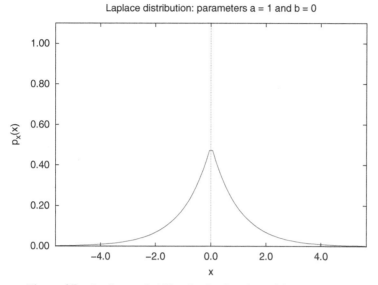

Figure 6.7 Laplace probability density function, with parameters $a = 1$ and $b = 0$.

Figure 6.7 illustrates the pdf for $a = 1$ and $b = 0$.

$$p_X(x) = \frac{a}{2} e^{-a|x-b|}. \tag{6.11}$$

For the Laplace distribution with zero mean, the probability that X is in the interval $(-\infty, x]$ is calculated as the integral of $p_X(x)$ in two parts,

For $-\infty < x \leq 0$, one obtains

$$P_X(x) = \int_{-\infty}^{x} p_X(x) \mathrm{d}x = \int_{-\infty}^{x} \frac{a}{2} e^{ax} \mathrm{d}x = \frac{1}{2} e^{ax} \mathrm{u}(-x). \tag{6.12}$$

On the other hand, for $0 < x < \infty$, one obtains

$$P_X(x) = \int_{-\infty}^{x} p_X(x) \mathrm{d}x = \int_{0}^{x} \frac{a}{2} e^{-ax} \mathrm{d}x = \left[\frac{1}{2} - \frac{1}{2} e^{-ax}\right] \mathrm{u}(x). \tag{6.13}$$

Therefore, the complete cumulative function, for the entire line, is given by

$$P_X(x) = \frac{1}{2} e^{ax} \mathrm{u}(-x) + \left[\frac{1}{2} - \frac{1}{2} e^{-ax}\right] \mathrm{u}(x). \tag{6.14}$$

Figure 6.8 illustrates the CDF for a Laplace distribution, considering the parameters $a = 1$ and $b = 0$.

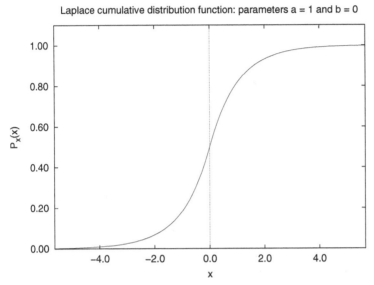

Figure 6.8 Laplace cumulative distribution function, with parameters $a = 1$ and $b = 0$.

6.2.1 Change of Variable Theorem

As discussed previously, given a random variable X on a probability triple $(\Omega, \mathcal{F}, P)$, the associated probability measure μ represents the distribution on the real space $\mathbb{R}$, with a Borel σ-algebra defined by (Rosenthal, 2000),

$$\mu(A) = P(X \in A) = P[X^{-1}(A)]. \tag{6.15}$$

The distribution defines completely the expected value of a random variable and, of course, of any function of it. Given a probability triple $(\Omega, \mathcal{F}, P)$, if X is a random variable that has distribution μ, then, for any measurable Borel function $f : \mathbb{R} \to \mathbb{R}$, one has

$$\int_\Omega f[X(\omega)] P(\mathrm{d}\omega) = \int_{-\infty}^{\infty} f(x) \mu(\mathrm{d}x), \tag{6.16}$$

provided that both sides are well-defined.

This means that the expected value of a function of the random variable, $f(X)$, with respect to the probability measure P on Ω is equal the expected value of the function with respect to the measure μ on $\mathbb{R}$.

Example: Consider that $f = I_A$, as defined previously, is the indicator function of a Borel set $A \in \mathbb{R}$. Therefore, one can write

$$\int_\Omega f[X(\omega)]P(\mathrm{d}\omega) = \int_\Omega I_{[X(\omega)\in A]}P(\mathrm{d}\omega) = P(X \in A).$$

But, as expected,

$$\int_{-\infty}^{\infty} f(x)\mu(\mathrm{d}x) = \int_{-\infty}^{\infty} I_{x\in A}\mu(\mathrm{d}x) = \mu(A) = P(X \in A),$$

and the equality holds in this case. The general case can be obtained considering that f is a linear combination of indicator functions.

6.3 Moments of a Random Variable

The moments of a random variable are statistics that reveal important information. They represent the variable's means, indicating trends and deviations. For example, considering the distribution of grades in a classroom, one can calculate the classroom's mean and also the deviation centered at this mean. In an election, it is possible to determine the voting trend. In a biological experiment, the confidence interval of the data can be calculated.

The mathematical expectation can be defined using the Lebesgue integral, particularized to the probabilistic measure space $(\Omega, \mathcal{F}, P)$. For such, some conditions of integrability are necessary for the random variable.

For the probability space $(\Omega, \mathcal{F}, P)$, consider the random variable $X : \Omega \mapsto \mathbb{R}$, which takes the event $\{\omega\} \in \mathcal{F}$ to the random variable $X(\omega)$. The mathematical expectation, also known as the expected value, of X, is defined as

$$\mathrm{E}[X] = \int_\Omega X dP = \int_\Omega X(\omega) dP(\omega). \qquad (6.17)$$

The knowledge of the correspondence, that defines the random variable as a function of a random event, is not always necessary to determine its expected value. It is sufficient to know the law, or measure, of probability of the random variable.

Considering that the measure used is the CDF, that is, $\mu(x) = P_X(x)$, which is positive and additive, then

$$\mathrm{E}[f(X)] = \int_\Omega f(X) dP = \int_\mathbb{R} f(x) dP_X(x). \qquad (6.18)$$

A function $f(x) : \mathbb{R} \mapsto \mathbb{R}$ is integrable relative to the measure $\mu(x) = P_X(x)$, if and only if $f(X)$ is integrable relative to the measure P.

The rightmost integral in Equation (6.18) is, in some texts, called the Stieltjes integral, or Riemann–Stieltjes integral, or even the Lebesgue integral relative to the probability measure $P_X(x)$.

To demonstrate the validity of Equation (6.18), a function $f = I_A$ can be considered, with $A \subset \mathcal{F}(\mathbb{R})$, and it can be verified that the equality results from having

$$f(X(\omega)) = I_A(X(\omega)) = \begin{cases} 1, & X(\omega) \in A, \text{ that is, if } \omega \in X^{-1}(A) \\ 0, & X(\omega) \notin A, \text{ that is, if } \omega \notin X^{-1}(A). \end{cases}$$

Therefore, $I_A(X) = I_{X^{-1}(A)}$, which implies

$$\int_\Omega f(X) dP = \int_\Omega I_{X^{-1}(A)} dP = P[X^{-1}(A)] = \mu(A) = \int_\mathbb{R} f(x) dP_X(x).$$

Let $f(X)$ be a random variable function, the expected value relative to X of the function $f(X)$, considering that the cumulative function $P_X(x)$ has the derivative $p_X(x)$, is given by

$$E[f(X)] = \int_{-\infty}^{\infty} f(x) p_X(x) dx. \quad (6.19)$$

6.3.1 Properties Associated to the Expected Value

As a result of the properties of the Lebesgue integral, the expected value operator, or mathematical expectation, has the following properties

$$E[\alpha X] = \alpha \cdot E[X], \quad (6.20)$$

for a given constant α.

$$E[X + Y] = E[X] + E[Y], \quad (6.21)$$

for any given random variables X and Y, and

$$E[XY] = E[X] \cdot E[Y] \quad (6.22)$$

if X and Y are independent random variables. X and Y are considered independent if the σ-algebras generated by them are independent, that is, for any Borel sets B and C in $\mathbb{R}$,

$$P(X^{-1}(B) \cap P(Y^{-1}(C)) = P(X^{-1}(B)) \times P(Y^{-1}(C)). \quad (6.23)$$

Many moments of the random variable X, defined in the probability space $(\mathbb{R}, \mathcal{B}((\mathbb{R}), P)$ have a special importance, and distinct physical interpretations for different domains. In a general manner, the moments are defined by the formula

$$\mathrm{E}[X^n] = \int_{-\infty}^{\infty} x^n p_X(x) \mathrm{d}x. \tag{6.24}$$

For a discrete probability distribution, of the type

$$p_X(x) = \sum_{k=-\infty}^{\infty} p_k \delta(x - x_k), \tag{6.25}$$

the expected value is calculated as

$$\begin{aligned}
\mathrm{E}[X^n] &= \int_{-\infty}^{\infty} x^n \sum_{k=-\infty}^{\infty} p_k \delta(x - x_k) dx \\
&= \sum_{k=-\infty}^{\infty} \int_{-\infty}^{\infty} p_k x^n \delta(x - x_k) dx \\
&= \sum_{k=-\infty}^{\infty} p_k x_k^n.
\end{aligned} \tag{6.26}$$

Note that the sampling property of the impulse function was used in the last step.

6.3.2 Definition of the Most Important Moments

The first moments are more well-known, usually being related to the means and physical measures. They are defined as:

- $m_1 = m_X = \mathrm{E}[X]$, arithmetic mean, mean value, average voltage value of a signal X, statistical average.
- $m_2 = P_X = \mathrm{E}[X^2]$, square mean, total power of a signal X.
- $m_3 = \mathrm{E}[X^3]$, moment or measure of the asymmetry (skewness) of the probability density function.
- $m_4 = \mathrm{E}[X^4]$, moment or measure of the flatness of pdf.

The variance, also known as the AC power of X in engineering, is an important statistical measure in Medicine, Engineering, Economics and other fields, being defined as follows

$$V[X] = \sigma_X^2 = E[(X-m_1)^2] = m_2 - m_1^2, \qquad (6.27)$$

in which the standard deviation, σ_X, is defined as the square root of the variance and indicates how much the variable deviates from its mean value.

The variance has the following properties, that can be verified using the definition,

$$V[\alpha X + \beta] = \alpha^2 V[X],$$

$$V\left[\sum_{i=1}^{N} X_i\right] = \sum_{i=1}^{N} V[X_i].$$

The coefficient of asymmetry, or skewness, is defined, for a random variable with expected value m_X and standard deviation σ, by using the corresponding moment (Magalhães, 2006).

$$\alpha_3 = \frac{E[(X - m_X)^3]}{\sigma_X^3}. \qquad (6.28)$$

The kurtosis coefficient, which measures the intensity of the peaks in the probability distribution of a random variable X, is defined as

$$\alpha_4 = \frac{E[(X - m_X)^4]}{\sigma_X^4}. \qquad (6.29)$$

The median of a random variable, m_D, is the measure of the position of the center of its distribution, and it divides the probability distribution into two parts with equal areas.

$$\int_{-\infty}^{m_D} p_X(x) dx = \frac{1}{2}. \qquad (6.30)$$

This measure is frequently used in the calculation of populations, specially in sociology and economics. It is equal to the mean if the pdf is symmetric.

Example: Consider the Gaussian, or Normal, distribution, illustrated in Figure 6.9 for some values of the mean and standard deviation, the formula for which is given by

$$p_X(x) = \frac{1}{\sigma_X \sqrt{2\pi}} e^{\frac{(x-m_X)^2}{2\sigma_X^2}}.$$

6.3 Moments of a Random Variable 131

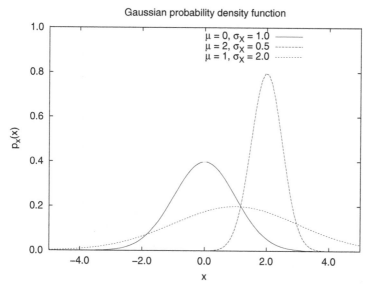

Figure 6.9 Gaussian probability density function, for three different values of the mean, μ, and standard variation, σ_X.

The mean value of the random variable is given by

$$E[X] = \int_{-\infty}^{\infty} x p_X(x) \, dx = \int_{-\infty}^{\infty} \frac{x}{\sigma_X \sqrt{2\pi}} e^{-\frac{(x-m_X)^2}{2\sigma_X^2}} \, dx.$$

However, it is known that the area under the curve of the normal function is unitary,

$$\int_{-\infty}^{\infty} \frac{1}{\sigma_X \sqrt{2\pi}} e^{-\frac{x^2}{2\sigma_X^2}} \, dx = 1.$$

Therefore, with a change of variables, it can be seen that

$$E[X] = \int_{-\infty}^{\infty} (y + m_X) \cdot \frac{1}{\sigma_X \sqrt{2\pi}} e^{-\frac{y^2}{2\sigma_X^2}} \, dy$$

$$= \frac{1}{\sigma_X \sqrt{2\pi}} \left[\int_{-\infty}^{\infty} y e^{-\frac{y^2}{2\sigma_X^2}} \, dy + m_X \int_{-\infty}^{\infty} e^{-\frac{y^2}{2\sigma_X^2}} \, dy \right]. \quad (6.31)$$

For the rightmost term, the first function to be integrated is odd in the considered interval, and the result of the integration is zero. The second

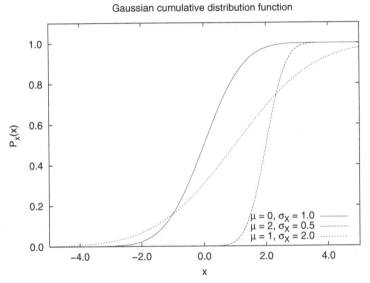

Figure 6.10 Gaussian cumulative distribution function, for three different values of the mean, μ, and standard deviation, σ_X.

integral is unitary, because it represents the area under the pdf. Therefore,

$$E[X] = m_X.$$

Since the Gaussian distribution is symmetric in relation to the mean, the mean and the median are the same.

The CDF is calculated integrating the pdf,

$$P_X(x) = \frac{1}{\sigma_X\sqrt{2\pi}} \int_{-\infty}^{x} e^{-\frac{(\alpha - m_X)^2}{2\sigma_X^2}} d\alpha.$$

The result is shown in Figure 6.10, for some mean and standard deviation values. ∎

6.4 Functions of Random Variables

The theory of functions of a random variable, also called the pdf transformation, is rich in applications in different areas of engineering. One application is illustrated in Figure 6.11, which shows the operation executed by the circuit described by the equation $Y = g(X)$. The circuit can be a linear

6.4 Functions of Random Variables

amplifier, an attenuator, a rectifier, a limiter, a logarithmic amplifier, or any other operation that does not involve energy storage devices, such as capacitors or inductors.

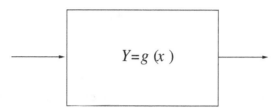

Figure 6.11 Application of transformation of pdf to the solution of a circuit, described by the equation $Y = g(X)$.

One method to calculate the probability distribution of the signal at the output of the circuit, assuming that it can be described by the associated random variable, is to use the cumulative function in terms of the measure of the Borel interval.

Consider an interval $(-\infty, y]$ and a probability measure P. It is known that the probability distribution $P_Y(y)$ of the output variable is the probability of the event $\{Y \leq y\}$, consisting of all the results ω such that $Y(\omega) = g[X(\omega)] \leq y$.

Therefore,

$$P_Y(y) = P\{Y \leq y\} = P\{g(X) \leq y\}. \tag{6.32}$$

There are a few necessary conditions so that $g(X)$ can be a random variable:

- The domain of $g(x)$ must include the image of the random variable X.
- The events $\{g(X) = \infty\}$ and $\{g(X) = -\infty\}$ must have zero probability.
- For $\{Y \leq y\}$ to be an event, the set of values of X for which $g(x) \leq y$ must be formed from the union and intersection of a countable number of Borel intervals.

If the requirements are met, and the function is bijective, the inverse $g^{-1}(y)$ can be applied in Equation (6.32), to obtain

$$P_Y(y) = P\{X \leq g^{-1}(y)\} = P_X(g^{-1}(y)). \tag{6.33}$$

Example: A linear system is governed by the equation $Y = AX$. Given the distribution $P_X(x)$ for the input variable, what is the distribution for the output variable? What is the pdf of the output random variable?

134 Random Variables

From the Formula 6.33, and assuming that the inverse function is given by $x = g^{-1}(y) = y/A$, then

$$P_Y(y) = P_X(y/A).$$

Differentiating the obtained result, and knowing that A can either be negative or positive, but not zero, yields

$$p_Y(y) = \frac{d}{dy}P_X(y/A) = \frac{p_X(y/A)}{|A|}.$$

For the case in which $A = 0$, which means there is no output signal, the probability measure is projected onto zero, and therefore the CDF is given by

$$P_Y(y) = u(y/A) = u(y),$$

and the pdf is

$$p_Y(y) = \delta(y). \blacksquare$$

The operation executed by the generic system shown in Figure 6.11 can also be seen, in the domain of the random variable, as illustrated in Figure 6.12. The probability distribution of the input variable is transformed by the system, generating a different distribution at the output.

Example: Assuming that X is a random variable, with a CDF given by

$$P_X(x) = P(X \leq x) = \frac{1}{(1+e^{-x})^\alpha},$$

in which $\alpha > 0$ is a given parameter. Consider a system that produces the following random variable at the output

$$Y = \ln(1 + e^{-X}).$$

Then, by definition,

$$P_Y(y) = P(Y \leq y) = P(\ln(1 + e^{-X}) \leq y) = P(X > -\ln(e^y - 1)).$$

The inequality can be evaluated in terms of the cumulative distribution of X,

$$\begin{aligned}
P_Y(y) &= 1 - P_X(-\ln(e^y - 1)) \\
&= 1 - \frac{1}{(1 + e^{\log(e^y - 1)})^\alpha} \\
&= 1 - \frac{1}{(1 + e^y - 1)^\alpha} \\
&= 1 - e^{-\alpha y}, \text{ for } y \geq 0.
\end{aligned} \qquad (6.34)$$

The last formula is the CDF of an exponential distribution, and the pdf is obtained by differentiating $P_Y(y)$, which gives

$$p_Y(y) = \frac{dP_Y(y)}{dy} = e^{-\alpha y}, \text{ for } y \geq 0. \blacksquare$$

Since the bijective mapping implies that the areas shown are equivalent, then

$$p_X(x)\,dx = p_Y(y)\,dy,$$

for which the solution is shown next, taking into account that the inversion of the curve of the system requires a change in the sign of the derivative,

$$p_Y(y) = \frac{p_X(x)}{|dy/dx|}, \quad x = g^{-1}(y).$$

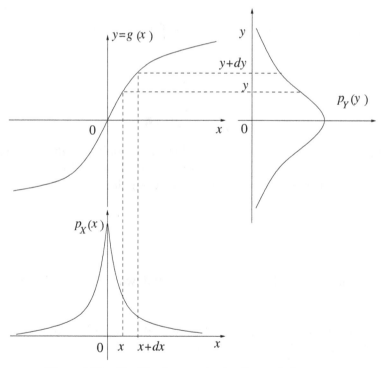

Figure 6.12 The bijective mapping implies equivalent areas.

The formula presented assumes the existence of the inverse of $f(\cdot)$ as well as its derivative at all points. Because the probability is always positive and dy/dx can be negative, the absolute value of the derivative must be taken. On the other hand, it is necessary to obtain $x = g^{-1}(y)$, because an answer is desired in terms of Y and the input is in terms of X.

Example: A random variable X is amplified by a linear system with positive gain G, that is,
$$Y = GX.$$
Obtain the pdf of Y, knowing that
$$p_X(x) = \alpha e^{-\alpha x} u(x).$$

$$\frac{dy}{dx} = G$$

$$x = \frac{y}{G}$$

Using the formula for the transformation of the pdf, yields
$$p_Y(y) = \frac{\alpha e^{-\alpha \frac{y}{G}} u(y/G)}{G}.$$

Therefore,
$$p_Y(y) = \frac{\alpha}{G} e^{-\frac{\alpha y}{G}} u(y). \blacksquare$$

Example: Consider an amplifier with a bias, B, modelled by the equation $Y = GX + B$, in which the gain, G, is positive. Note that the function is not linear, it is rather known as an affine function. A function is linear if it is:

1. homogeneous, that is, $\alpha x \to \alpha y$;

2. additive, that is, $x_1 + x_2 \to y_1 + y_2$.

The solution for the problem is obtained by following the steps illustrated in Figure 6.13:

1. Obtain the differential gain, or differentiate the output in relation to the input. In this case, $\frac{dy}{dx} = G$, $G > 0$;

2. Obtain the inverse of the function that relates the input to the output, that is, $x = \frac{y-B}{G}$;

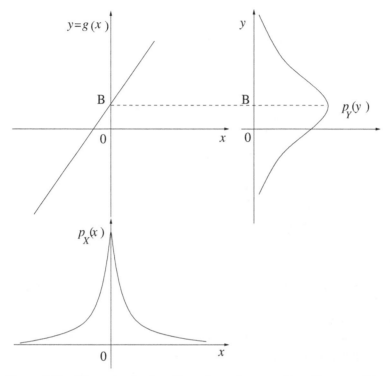

Figure 6.13 Illustration for the pdf transformation caused by a linear amplifier.

3. Apply the formula $p_Y(y) = \frac{p_X(x)}{|dy/dx|} = \frac{p_X((y-B)/G)}{G}$. ∎

Example: Substituting the Gaussian distribution from Figure 6.9 in the previous example, in which $Y = GX + B$, results in

$$p_X(x) = \frac{1}{\sigma_X\sqrt{2\pi}} e^{-\frac{x^2}{2\sigma_X^2}}.$$

$$p_Y(y) = \frac{p_X(x)}{|\frac{dy}{dx}|},$$

$$y = Gx + B \Rightarrow x = \frac{y-B}{G}, \qquad \frac{dy}{dx} = G > 0.$$

Therefore,

$$p_Y(y) = \frac{p_X\left(\frac{y-B}{G}\right)}{G},$$

$$p_Y(y) = \frac{1}{\sigma_X\sqrt{2\pi}G} e^{-\frac{\left(\frac{y-B}{G}\right)^2}{2\sigma_X^2}},$$

Using $\sigma_X G = \sigma_Y$, yields

$$p_Y(y) = \frac{1}{\sigma_Y\sqrt{2\pi}} e^{-\frac{(y-B)^2}{2\sigma_Y^2}}. \blacksquare$$

Considering X as an electrical signal, the following relationships can be verified:

- The total power of the input signal is $P_X = \sigma_X^2$.
- The average power, also called direct current (DC) power, of the input signal is $P_{DC} = m_X^2 = 0$.
- The variance is the AC power, given by $P_{AC} = V[X] = \sigma_X^2$.

For the output signal:

- The variance is given by $P_{AC} = V[Y] = \sigma_Y^2 = G^2 \sigma_X^2$, which results in an effective value $V_{RMS} = \sigma_Y$.
- The mean level of the output signal is $P_{DC} = m_Y^2 = B^2$.
- The total power of the output signal is $P_Y = G^2 \sigma_X^2 + B^2$. It is a function of the square of the gain and of the square of the bias voltage.

6.4.1 General Formula for Transformation

The general formula for the calculation of the pdf transformation, which includes the cases where the function has partial only inverses, or does not have an inverse for the entire interval, can be obtained with use of the impulse function, created by Paul Adrien Maurice Dirac (1902–1984), an English electrical engineer and mathematician, who graduated from Bristol University.

It is important to take into account that the function has inverses only in the intervals i, and that they are given by $f_i^{-1}(y)$, while it is constant for the intervals j, and hence the function does not have an inverse in those intervals. Therefore, the output presents impulses for the points y_j, with areas a_j. Then, the pdf of the system response is given by

$$p_Y(y) = \sum_i \frac{p_X(x_i)}{|dy/dx_i|}\Big|_{x_i=f_i^{-1}(y)} + \sum_j a_j \delta(y-y_j). \tag{6.35}$$

6.4 Functions of Random Variables

The preceding formula permits the computation of the output probability function for any reasonable transformation $y = f(x)$. Of course, there are functions, such as Dedekind's, which requires the use of the Lebesgue integral.

Example: If it is known that $Y = X^2$, and that $p_X(x) = \frac{1}{\sqrt{2\pi}} e^{\frac{-x^2}{2}}$, $-\infty < x < \infty$, determine $p_Y(y)$.

$$p_Y(y) = \frac{p_X(x)}{|dy/dx|} \bigg|_{x=f^{-1}(y)}.$$

$$\left|\frac{dy}{dx}\right| = |2x|.$$

Since the function $y = x^2$ does not admit an inverse, the problem must be solved in parts and the results added.

Region I: $x = \sqrt{y}$,

$$p_{Y1}(y) = \frac{1}{2\sqrt{2\pi y}} e^{-\frac{y}{2}}.$$

Region II: $\rightarrow x = -\sqrt{y}$,

$$p_{Y2}(y) = \frac{1}{2\sqrt{2\pi y}} e^{-\frac{y}{2}}.$$

Now,

$$p_Y(y) = p_{Y1}(y) + p_{Y2}(y),$$

which results in

$$p_Y(y) = \frac{1}{\sqrt{2\pi y}} e^{-\frac{y}{2}}, \ 0 < y < \infty. \blacksquare$$

This pdf is called the chi-square distribution and has the following general formula

$$P_X(x) = \frac{x^{n/2-1} e^{-x^2}}{2^{n/2} \Gamma(n/2)}, \ 0 < x < \infty, \tag{6.36}$$

in which

$$\Gamma(n) = \int_0^\infty x^{n-1} e^{-x} dx, \ n > 0. \tag{6.37}$$

140 *Random Variables*

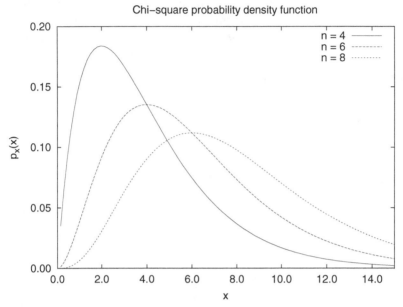

Figure 6.14 Chi-square probability density function, for three different values of the parameter n.

The distribution is shown in Figure 6.14. It is typically used in hypothesis testing for biometrics to find, for example the dispersion value of two nominal variables, evaluating the existing association between qualitative variables.

Two groups are said to behave in a similar manner if the difference between the frequencies observed and the frequencies expected in each category is small. The chi-square CDF is illustrated in Figure 6.15.

Example: A signal with a Gaussian distribution is applied to a full-wave rectifier circuit.

Once again considering the signal has the distribution,

$$p_X(x) = \frac{1}{\sigma_X \sqrt{2\pi}} e^{-\frac{x^2}{2\sigma_X^2}},$$

a qualitative analysis, illustrated in Figure 6.16, indicates that the output of the rectifier is folded, so that the output signal has a distribution that occupies half the domain, but has double the amplitude so as to maintain the unitary area underneath the curve.

6.4 Functions of Random Variables

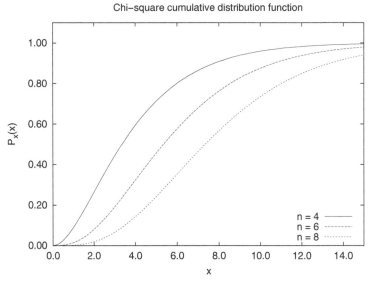

Figure 6.15 Chi-square cumulative distribution function, for three different values of the parameter n.

$$p_Y(y) = \frac{2}{\sigma_X \sqrt{2\pi}} e^{-\frac{y^2}{2\sigma_X^2}} u(y).$$

In addition to this, the Root Mean Square (RMS) value of the signal is reduced, because it is calculated as a function of the width of the pdf, which is reduced during the rectification. ■

Example: A method for building a random process is presented in the following. Consider the random variable ϕ with uniform distribution in the interval $[0, 2\pi]$, shown in Figure 6.17.

The function $Y = V \cos \phi$ does not have an inverse for every point, therefore, it must be divided into parts to obtain the inverses.

$$p_Y(y) = \frac{p_\phi(\phi)}{|\mathrm{d}y/\mathrm{d}\phi|}, \qquad y = V \cos \phi \Rightarrow \phi = \cos^{-1}(y/V),$$

$$\frac{\mathrm{d}y}{\mathrm{d}\phi} = -V \operatorname{sen} \phi = -V\sqrt{1 - (y/V)^2} = -\sqrt{V^2 - y^2},$$

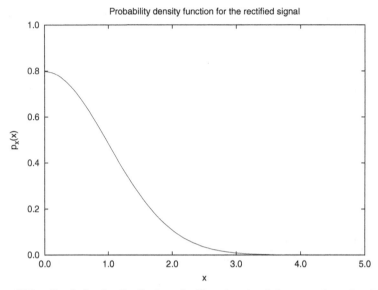

Figure 6.16 Result for the distribution of a Gaussian signal that goes through a full-wave rectification.

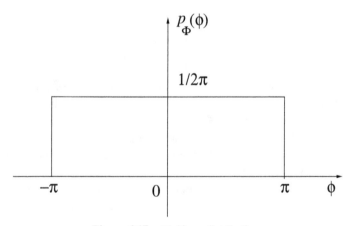

Figure 6.17 Uniform distribution.

$$p_Y(y) = 2 \cdot \left[\frac{1}{2\pi} \cdot \frac{1}{\sqrt{V^2 - y^2}} \right] = \frac{1}{\pi \sqrt{V^2 - y^2}}.$$

The result is illustrated in Figure 6.18. It can be noticed that the distribution obtained represents the probability of the sine wave, for which the more

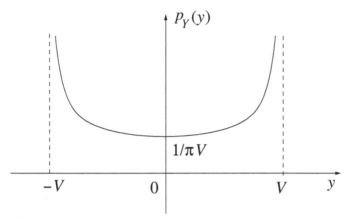

Figure 6.18 The distribution function obtained from the mapping $y = V \cos \phi$.

extreme values have a greater probability of occurrence. A random process can be built merely by including the variable time in the discussed equation $Y = V \cos(\omega t + \phi)$. ∎

Furthermore, with regard to the pdf transformation, also called the function of a random variable, there are two situations of particular interest involving the uniform distribution:

First situation: transformation of a given pdf into a uniform pdf.

In this case, the output distribution is given by

$$p_y(y) = \frac{p_X(x)}{|\frac{dy}{dx}|}, \quad x = f_{-1}(y), \quad \frac{dy}{dx} \geq 0,$$

$$p_y(y) = \frac{p_X(x)}{dy/dx} \Rightarrow 1 = \frac{p_X(x)}{dy/dx}, \quad 0 \leq y \leq 1.$$

Therefore,

$$p_X(x) = \frac{dy}{dx} \Rightarrow dy = p_X(x)\,dx.$$

Then, integrating the differential in y, the desired function is obtained

$$f(x) = \int_{-\infty}^{y} dy = \int_{-\infty}^{x} p_X(x)\,dx = P_X(x).$$

The operation is useful, for example, for the analysis of signal compression and expansion systems, or for the study of non-uniform quantizers, which are used in telephony systems.

144 Random Variables

Second situation: transformation of a uniform distribution into another given pdf.

Consider the calculation of the pdf of the output of the system,

$$p_Y(y) = \frac{p_X(x)}{dy/dx}, \quad p_X(x) = 1, \quad 0 \le x \le 1.$$

Therefore,

$$p_y(y) = \frac{1}{dy/dx} \Rightarrow \frac{dy}{dx} = \frac{1}{p_y(y)}, \quad y = g(x).$$

Solving the equation to obtain y, it is seen that the function required to transform the uniform distribution into another given distribution is the inverse of the CDF of the input, as follows

$$y = \int_{-\infty}^{x} \frac{1}{p_Y(g(x))} dy \Rightarrow y = P_X^{-1}(x).$$

This operation is useful to generate the random variables of any distribution in a computer, using a variable with uniform distribution, which is typically found in any computational language.

6.5 Discrete Distributions

The discrete probability distributions can be written, with the set of real numbers $\mathbb{R}$, using the generalized impulse function, or the Dirac Delta function.

The pdf for a random variable with Bernoulli distribution is given by

$$p_X(x) = p\,\delta(x+a) + q\,\delta(x-b), \tag{6.38}$$

illustrated in Figure 6.19, and its CDF is

$$P_X(x) = p\,u(x+a) + q\,u(x-b), \tag{6.39}$$

which is illustrated in Figure 6.20, in which $p + q = 1$.

In the case where levels a and b are equal, and the associated probabilities, p and q, are equal, then

$$p_X(x) = \frac{1}{2}\left[\delta(x+a) + \delta(x-a)\right]. \tag{6.40}$$

6.5 Discrete Distributions

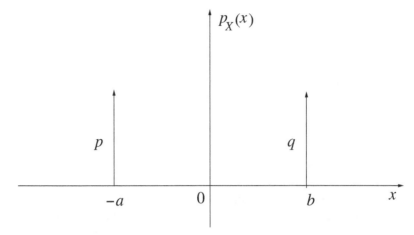

Figure 6.19 Bernoulli probability density function.

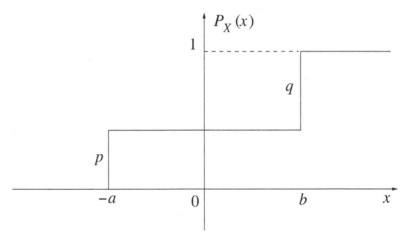

Figure 6.20 Bernoulli cumulative distribution function.

and its CDF is

$$P_X(x) = \frac{1}{2}\left[\mathrm{u}(x+a) + \mathrm{u}(x-a)\right]. \tag{6.41}$$

This distribution usually models data transmission in a communications system.

The binomial pdf is a discrete distribution of the number of successes in a sequence of n independent Bernoulli trials. In each trial, the probability of success, p, is the same.

$$p_X(x) = \sum_{k=0}^{\infty} \binom{n}{k} p^k q^{n-k} \delta(x-k), \qquad (6.42)$$

and the CDF is written as

$$P_X(x) = \sum_{k=0}^{\infty} \binom{n}{k} p^k q^{n-k} \mathrm{u}(x-k). \qquad (6.43)$$

The expected value of the binomial random variable is $\mathrm{E}[X] = np$ and its variance is given by $\mathrm{V}[X] = np(1-p)$.

The geometric distribution, which has its pdf as follows, models the arrival of packets in a computer network, when the arrival and service rates are constant.

$$p_X(x) = \sum_{k=0}^{\infty} (1-\rho)\rho^k \delta(x-k). \qquad (6.44)$$

The geometric pdf is shown in Figure 6.21. The heights of the impulses are merely illustrative, and only indicate the relative areas.

The geometric CDF is given by

$$P_X(x) = \sum_{k=0}^{\infty} (1-\rho)\rho^k \mathrm{u}(x-k). \qquad (6.45)$$

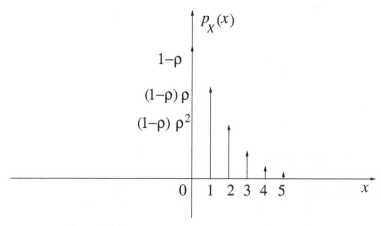

Figure 6.21 Geometric probability density function.

6.5 Discrete Distributions   147

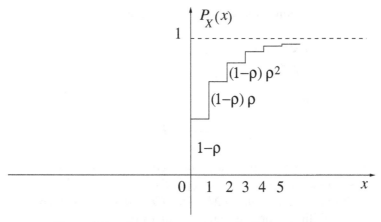

Figure 6.22   Geometric cumulative distribution function.

The geometric CDF is illustrated in Figure 6.22.

Example: Calculate the mean value of a random variable with a geometric distribution.

According to Equation (6.26), the mean value of a discrete random variable can be calculated by

$$E[X] = \sum_{k=0}^{\infty} p_k x_k$$
$$= \sum_{k=0}^{\infty} k(1-\rho)\rho^k$$
$$= (1-\rho) \sum_{k=0}^{\infty} k\rho^k.$$

Because the geometric series converges for $\rho < 1$,

$$\sum_{k=0}^{\infty} \rho^k = \frac{1}{1-\rho},$$

and

$$\frac{d}{d\rho}\left(\frac{1}{1-\rho}\right) = \frac{1}{(1-\rho)^2},$$

and the derivative of a sum is the sum of the derivatives,

$$\frac{d}{d\rho}\left(\sum_{k=0}^{\infty}\rho^k\right) = \sum_{k=0}^{\infty}k\rho^{k-1} = \frac{1}{\rho}\sum_{k=0}^{\infty}k\rho^k,$$

then the expected value is

$$\mathrm{E}[X] = \frac{\rho}{1-\rho}.\blacksquare$$

The Poisson distribution, illustrated in Figure 6.23, was discovered by Siméon Denis Poisson (1781–1840) and published in his work *Recherches sur la Probabilité des Jugements en Matières Criminelles et Matière Civile*, in which his probability theory also appears, in 1838.

The Poisson pdf is given by

$$p_X(x) = \sum_{k=0}^{\infty}\frac{e^{-\lambda}\lambda^k}{k!}\delta(x-k), \qquad (6.46)$$

and the CDF is

$$P_X(x) = \sum_{k=0}^{\infty}\frac{e^{-\lambda}\lambda^k}{k!}u(x-k). \qquad (6.47)$$

The Poisson CDF is illustrated in Figure 6.24. This distribution is commonly used to describe situations like particle emissions from radioactive

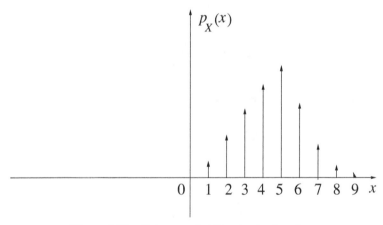

Figure 6.23   Poisson probability density function.

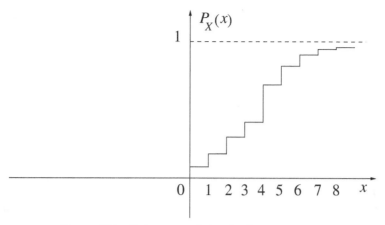

Figure 6.24 Poisson cumulative distribution function.

materials, telephone traffic, and even the formation of lines in banks. The expected value of a Poisson variable and its variance are equal to the distribution parameter λ.

An important aspect to consider with discrete probability distributions is that, for a finite, or countable, number of points, it is possible to assign nonzero probability values for all the points in the set. One may notice that, for a non-countable set of points, that is, for continuous distributions, it is impossible to assign nonzero probabilities to all points of the set. That is why the pdf is used when dealing with continuous variables, instead of sequences of probabilities, which results in the need to use the impulse function to model the discrete case.

6.6 Characteristic Function

The characteristic function is the Fourier transform of the pdf. It was created by Dirac as a way to solve the problem of the distribution of a sum of random variables.

$$P_X(\omega) = \mathrm{E}[e^{-j\omega X}] = \int_{-\infty}^{\infty} e^{-j\omega x} p_X(x)\,\mathrm{d}x. \qquad (6.48)$$

For discrete probability distributions, like Equation (6.25), the characteristic function can be calculated as

$$P_X(\omega) = \int_{-\infty}^{\infty} e^{-j\omega x} \sum_{k=-\infty}^{\infty} p_k \delta(x - x_k) \, dx.$$

$$= \sum_{k=-\infty}^{\infty} p_k \int_{-\infty}^{\infty} e^{-j\omega x} \delta(x - x_k) \, dx.$$

$$= \sum_{k=-\infty}^{\infty} p_k e^{-j\omega x_k}, \qquad (6.49)$$

which is the discrete Fourier transform of the distribution, or sequence of probabilities $\{p_k\}$.

The characteristic function is obtained by making $g(X) = e^{-j\omega X}$ in the general expression

$$E[g(X)] = \int_{-\infty}^{\infty} g(x) p_X(x) \, dx.$$

If the random variable X is given in volts, then $p_X(x)$ is given in [V^{-1}] and $P_X(\omega)$, the characteristic function, is always dimensionless. The unit for ω is the inverse of the unit for the random variable.

The characteristic function can also be used to find the moments of a random variable, provided the n-th moment exists. Consider the first and second order moments of the random variable X, given by

$$E[X] = \int_{-\infty}^{\infty} x p_X(x) \, dx$$

and

$$E[X^2] = \int_{-\infty}^{\infty} x^2 p_X(x) \, dx.$$

Differentiating Equation (6.48),

$$\frac{\partial}{\partial \omega} P_X(\omega) = \frac{\partial}{\partial \omega} \int_{-\infty}^{\infty} e^{-j\omega x} p_X(x) \, dx = \int_{-\infty}^{\infty} (-j) x e^{j\omega x} p_X(x) \, dx.$$

Making $\omega = 0$ in the equation yields

$$\frac{1}{-j} \cdot \frac{\partial}{\partial \omega} P_X(\omega) \bigg|_{\omega=0} = \int_{-\infty}^{\infty} x p_X(x) \, dx = E[X],$$

which is the expected value of X.

The second derivative produces the second moment,

$$\frac{1}{(-j)^2} \cdot \frac{\partial^2}{\partial \omega^2} P_X(\omega)\bigg|_{\omega=0} = \int_{-\infty}^{\infty} x^2 p_X(x)\, \mathrm{d}x = \mathrm{E}[X^2].$$

Therefore, all the moments of the random variable X can be obtained from the characteristic functions, through successive differentiation

$$\mathrm{E}[X^n] = \frac{1}{(-j)^n} \cdot \frac{\partial^n}{\partial \omega^n} P_X(\omega)\bigg|_{\omega=0}. \tag{6.50}$$

Example: Consider the distribution $p_X(x) = \alpha\, e^{-\alpha x}\, u(x)$. Its characteristic function is calculated as follows

$$P_X(\omega) = \int_{-\infty}^{\infty} e^{-j\omega x} p_X(x)\, \mathrm{d}x = \int_0^{\infty} e^{-j\omega x} \cdot \alpha \cdot e^{-\alpha x}\, \mathrm{d}x = \alpha \int_0^{\infty} e^{-x(j\omega+\alpha)}\, \mathrm{d}x,$$

$$P_X(\omega) = \frac{\alpha}{(\alpha + j\omega)} e^{x(j\omega+\alpha)}\bigg|_0^{\infty} = \frac{\alpha}{(\alpha + j\omega)}.$$

The first moment, or the mean of the random variable, is determined by calculating the first derivative of the characteristic function

$$\mathrm{E}[X] = \frac{1}{-j} \cdot \frac{\partial P_X(\omega)}{\partial \omega}\bigg|_{\omega=0} = \frac{1}{-j} \left[\frac{-\alpha \cdot (j)}{(\alpha + j\omega)^2}\right]\bigg|_{\omega=0} = \frac{1}{\alpha}.$$

The result can be verified computing the mean by the usual formula. ∎

6.7 Conditional Distribution

A probability distribution can be conditioned, by using Bayes formula for the conditional probability between sets, and defining the sets accordingly

$$P(A|B) = \frac{P(A \cap B)}{P(B)}.$$

Set A is defined as $A = \{X \leq x\}$, to make it possible to calculate the CDF

$$P\{X \leq x\} = P_X(x).$$

Set B can be defined using any line segment obtained from the Borel family, which is the family of all subsets from the real set $\mathbb{R}$ that can be represented by semi-open line segments, that is, $B = \{a < X \leq b\}$, as shown in Figure 6.25.

The probability measure of this set is given by

$$P\{a < X \leq b\} = P_X(b) - P_X(a).$$

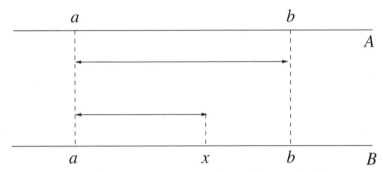

Figure 6.25 Line segments used in conditional probability.

Example: Suppose that $B = (-\infty, a]$, as indicated in Figure 6.26. The conditional probability is then given by

$$P\{X \leq x | B\} = \frac{P\{X \leq x, -\infty < X \leq a\}}{P\{-\infty < X \leq a\}}.$$

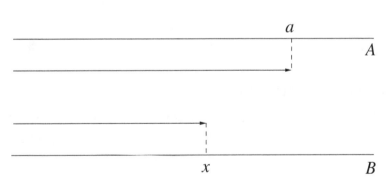

Figure 6.26 Line segments used in the example.

Considering that the intersection of the segments for $X \leq a$ is the set $\{X \leq x\}$ and that the intersection for $X > a$ is $\{X \leq a\}$, then

$$P_X(x|B) = P\{X \leq x | B\} = \begin{cases} \frac{P_X(x)}{P_X(a)}, & x \leq a \\ \frac{P_X(a)}{P_X(a)} = 1, & x > a \end{cases}.$$

6.7 Conditional Distribution

The conditional pdf is obtained by differentiating the conditional CDF

$$p_X(x|B) = \frac{d}{dx}P_X(x|B) = \begin{cases} \frac{p_X(x)}{P_X(a)}, & X \leq a \\ 0, & X > a \end{cases}.$$

The two functions are in Figure 6.27, in which the dashed line indicates the conditional distributions. One may notice that conditioning a CDF is equivalent to providing information about the random variable.

The knowledge that the variable is, *a priori*, in a given interval elevates the CDF, as well as the pdf. This can easily be seen, because the denominator is always less than one, unless the interval is the set of real numbers itself. ■

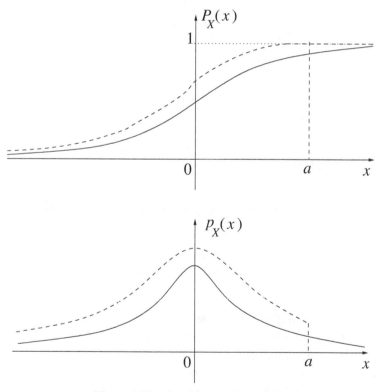

Figure 6.27 Conditional pdf and CDF.

Example: Consider an interval $B = [a, b]$. The conditional distribution is obtained as was done in the previous calculations, and is illustrated in

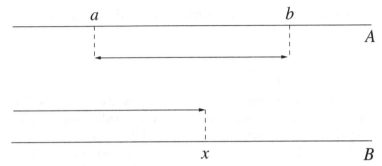

Figure 6.28 Line segments used to calculate the conditional probability.

Figure 6.28,
$$P_X(x|B) = \frac{P\{X \leq x, a < X \leq b\}}{P\{a < X \leq b\}},$$

in which, by definition,
$$P\{a < X \leq b\} = P_X(b) - P_X(a).$$

The conditional CDF is given by
$$P_X(x|B) = \begin{cases} 0 & , \quad X \leq a \\ \frac{P_X(x) - P_X(a)}{P_X(b) - P_X(a)} & , \quad a < X \leq b \\ 1 & , \quad X > b \end{cases}$$

and the conditional pdf can be immediately obtained as
$$P_X(x|B) = \begin{cases} 0 & , \quad X \leq a \\ \frac{p_X(x)}{P_X(b) - P_X(a)} & , \quad a < X \leq b \\ 0 & , \quad X > b. \end{cases}$$

Both functions are shown in Figure 6.29, in which the dashed line indicates the conditional distributions. ■

Example: Consider Figure 6.30, that represents a channel with additive noise, commonly used to model communications systems.

The transmitted signal initially assumes value A. The additive signal, X, that can be the noise, is usually independent of the transmitted signal.

This problem can be solved using conditional probability, or by transformation of the pdf. Using the second approach leads to

$$p_Y(y) = \frac{p_X(x)}{|dy/dx|}, \quad |dy/dx| = 1, \quad x = y - A$$

6.7 Conditional Distribution

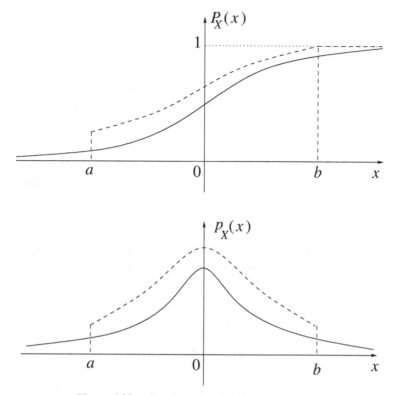

Figure 6.29 Conditional probability distributions.

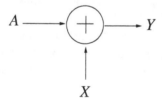

Figure 6.30 Additive noise channel, $Y = A + X$.

$$p_Y(y) = p_X(y - A)$$

Using conditional probability yields

$$P_Y(y|A) = \frac{P_Y(y, A)}{P(A)} = \frac{P_X(y - A) \cdot P(A)}{P(A)}$$

Simplifying the expression and calculating its derivative yields the result previously obtained.

For the case where the input is $-A$ or A, then

$$p_Y(y) = \frac{1}{2}p_X(y+A) + \frac{1}{2}p_X(y-A).$$

Note that the noise causes an intersection between the pdf curves. This interference is responsible for the detection errors in the received signal. The error probability is obtained using Bayes Rule,

$$P_e = P(e|A)P(A) + P(e|-A)P(-A),$$

in which $P(e|A)$ is the error probability given that the pulse with value A was transmitted, and $P(e|-A)$ represents the error probability in the case where $-A$ was transmitted.

Increasing the signal power improves detection, as it causes a separation between the conditional pdf curves. The transmitted signal power is A^2. ■

Example: For the specific case of an Additive White Gaussian Noise (AWGN) channel, the formulas become

$$p_N(n) = \frac{1}{\sigma_N\sqrt{2\pi}} e^{-\frac{N^2}{2\sigma_N^2}}$$

and

$$p_Y(y) = \frac{1}{2}\frac{1}{\sigma_N\sqrt{2\pi}} e^{-\frac{(n+A)^2}{2\sigma_N^2}} + \frac{1}{2} \cdot \frac{1}{\sigma_N\sqrt{2\pi}} e^{-\frac{(n-A)^2}{2\sigma_N^2}}.\ ■$$

Gaussian noise is common in communications systems, in which it typically occurs as thermal noise, in control systems, in which it appears as the estimation error of the controller, and in power systems, in which it models the effects of harmonics caused by thyristor-based equipment. The Gaussian error is also important to characterize population statistics.

6.8 Useful Distributions and Applications

A well-accepted model used for signal amplitude statistics in an environment subject to fading, uses the Rayleigh distribution, named after John William Strutt (1842–1919), English mathematician and physicist, also known as Baron of Rayleigh who has done extensive research on ondulatory phenomena, also known as Baron of Rayleigh (Kennedy, 1969).

This distribution, shown in Figure 6.31, represents the effect of multiple signals, reflected and refracted, when detected by the receiver, with no direct line of sight, or direct ray (Lecours et al., 1988).

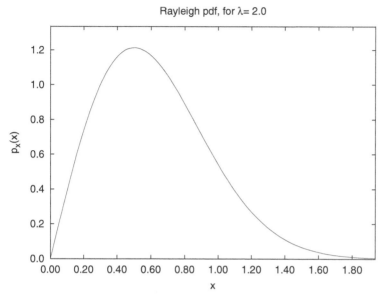

Figure 6.31 Rayleigh probability density function.

The Rayleigh pdf, illustrated in Figure 6.31 for different parameter values, is given by (Proakis, 1990)

$$p_X(x) = \frac{x}{\sigma^2} e^{-\frac{x^2}{2\sigma^2}}, u(x) \qquad (6.51)$$

with mean $E[X] = \sigma\sqrt{\pi/2}$ and variance $V[X] = (2-\pi)\frac{\sigma^2}{2}$. It's CDF is shown in Figure 6.32.

In this case, the phase distribution can be considered uniform for the interval $(0, 2\pi)$. It can be observed that a distribution equivalent to Rayleigh can be obtained with only six waves, with independently distributed phases (Schwartz et al., 1966).

If a strong main component is considered, that is, if the previously described scenario has a line of sight between transmitter and receiver, in addition to the multipath components, the Rice distribution, named after a pioneer in the fields of Communication and Information Theory, Stephen Oswald Rice (1907–1986), can be used to describe the fast variations in the signal envelope. This line of sight component reduces the variance in the signal amplitude, as it increases in comparison to the multipath components

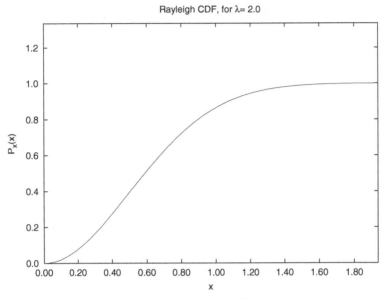

Figure 6.32 Rayleigh cumulative distribution function.

(Lecours et al., 1988) (Rappaport, 1989). The Rician distribution is given by

$$p_X(x) = \frac{x}{\sigma^2} e^{-\frac{x^2+A^2}{2\sigma^2}} I_0\left(\frac{xA}{\sigma^2}\right) u(x), \tag{6.52}$$

in which $I_0(\cdot)$ is the zero-order Bessel function, written as

$$I_0(x) = \frac{1}{\pi} \int_0^\pi e^{x \cos\theta} d\theta. \tag{6.53}$$

and A is the signal amplitude. The variance, for a unitary mean, is $V[X] = A^2 + 2\sigma^2 + 1$. And the expected value is given by

$$E[X] = e^{-\frac{A^2}{4\sigma^2}} \sqrt{\frac{\pi}{2}} \sigma \left[\left(1 + \frac{A^2}{2\sigma^2}\right) I_0\left(\frac{A^2}{4\sigma^2}\right) + \frac{A^2}{2\sigma^2} I_1\left(\frac{A^2}{4\sigma^2}\right) \right] \tag{6.54}$$

in which $I_1(\cdot)$ is the modified first-order Bessel function, written as

$$I_1(x) = \frac{1}{\pi} \int_0^\pi e^{x \cos\theta} \cos(\theta) d\theta. \tag{6.55}$$

6.8 Useful Distributions and Applications

The term $A^2/2\sigma^2$ is a measure of the fading statistics. As $A^2/2\sigma^2$ increases, the effect of the multiplicative noise, or fading, becomes less important. The pdf of the signal becomes more concentrated around the main component. The remaining disturbances are perceived as phase fluctuations.

On the other hand, the signal weakening can cause the main component to remain undetected among the multipath components, which results in the Rayleigh model. And increasing A in relation to the standard deviation, σ, makes the statistics converge to a Gaussian distribution with mean A (Schwartz, 1970).

Another distribution that is useful when modelling multipath fading scenarios is the Nakagami pdf. This distribution can be applied in cases where there is a random superposition of random vectorial components (Shepherd, 1988). The Nakagami distribution can be expressed as

$$p_X(x) = \frac{2m^m x^{2m-1}}{\Gamma(m)\Omega^m} e^{-\frac{mx^2}{\Omega}} u(x), \qquad (6.56)$$

in which $\Omega = P_X$ is the mean power of the received signal, $m = \Omega^2/\mathrm{E}[(X - \mathrm{E}[X])^2]$ represents the inverse of the normalized variance of X^2 and $\Gamma(\cdot)$ is the Gamma function. The parameter m is known as the distribution modelling factor and cannot be less than $1/2$.

It can be shown that the Nakagami pdf represents a more general expression, that covers many other distributions. For example, for $m = 1$ the Rayleigh distribution is obtained.

The lognormal distribution is used to model the effect of certain topography patterns on the transmitted signal, that appear because of the non-homogeneity of the channel or due to transmission in very obstructed or congested environments (Hashemi, 1991; Rappaport, 1989).

Certain natural growth processes are driven by the accumulation of small percentage changes which become additive on a logarithm scale. The distribution of the resulting accumulated changes is well approximated by a lognormal, under appropriate conditions. The lognormal distribution, which can also model communication channels in industrial environments, is represented by

$$p_R(r) = \frac{1}{\sigma r \sqrt{2\pi}} e^{-\frac{(\log r - \mu)^2}{2\sigma^2}} u(r) \qquad (6.57)$$

which has mean value $\mathrm{E}[X] = e^{\sigma^4/2 + \mu}$, and variance $V[X] = e^{\sigma^4 + \mu}(e^{\sigma^4} - 1)$.

It can be obtained directly from the Gaussian distribution by the use of an appropriate transformation. Its pdf is shown in Figure 6.33, for $\mu = 1$ and $\sigma = 0, 5$.

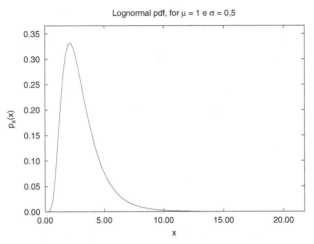

Figure 6.33 Lognormal pdf.

The lognormal distribution is used to model the time periods to repair a maintainable system, In reliability analysis. In wireless communication, it can model the random obstruction of radio signals due to large buildings and hills. This phenomenon is called shadowing. The lognormal distribution also mode the interference distribution caused bay a sudden outbreak of information in the network (Cordeiro et al., 2021).

The file size distribution of publicly available audio and video data files, such as MIME types, follows a lognormal distribution. The lognormal distribution is as a good statistical model to represent the amount of traffic per unit time in computer networks and Internet traffic analysis, The length of comments posted in Internet discussion forums, as well as, the time a user spends on online articles follow a lognormal distribution.

The CDF of the lognormal distribution is illustrated in Figure 6.33. Note that the function increases faster than the Gaussian CDF.

The von Mises distribution was introduced in 1918, by Richard Edler von Mises (1883–1953), an Austrian mechanical engineer and mathematician, for the study of deviations in atomic weights in relation to integer values (Mises, 1918). Recently, it has an important role in the modeling and statistical analysis of angular variables. Consider the random variable

6.8 Useful Distributions and Applications

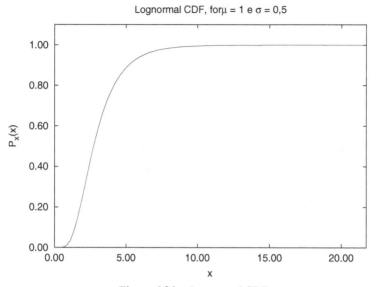

Figure 6.34 Lognormal CDF.

Θ representing the angle of arrival of a certain signal, from a multipath component (scattered and specular). The von Mises distribution for the scattered component of Θ is given by

$$p_\Theta(\theta) = \frac{1}{2\pi I_0(\kappa)} e^{\kappa \cos(\theta - \theta_M)}, \quad \theta \in [-\pi, \pi), \tag{6.58}$$

in which $I_0(\cdot)$ is the modified zero-order Bessel function and $\theta_M \in [-\pi, \pi)$ represents the mean direction for the angle of arrival of the scattered components and $\kappa \geq 0$ controls the width of the arrival angle.

The distribution discovered by the Italian civil engineer and economist Vilfredo Federico Damaso Pareto (1848–1923) is defined as (Pareto, 1964)

$$p(x) = \frac{\alpha \beta^\alpha}{x^{\alpha+1}}, \quad \alpha, \beta > 0. \tag{6.59}$$

The CDF is

$$P(x) = 1 - \left(\frac{\beta}{x}\right)^\alpha, \quad x > \beta. \tag{6.60}$$

The first moment, or expected value is given by

$$m_1 = E[x] = \frac{\alpha \beta}{(\alpha - 1)}, \quad \alpha > 1. \tag{6.61}$$

The distribution is named after Waloddi Weibull (1887–1979), a Swedish engineer and mathematician, and is useful to describe a particle size distribution. It is defined as Weibull (1951)

$$p(x) = \alpha x^{\alpha-1} \beta^{-\alpha} e^{-\left(\frac{x}{\beta}\right)^\alpha}, \quad \alpha, \beta > 0. \tag{6.62}$$

The CDF is
$$P(x) = 1 - e^{-\left(\frac{x}{\beta}\right)^\alpha}. \tag{6.63}$$

And the first moment, or expected value, is written as

$$m_1 = E[x] = \beta \Gamma\left(1 + \frac{1}{\alpha}\right), \tag{6.64}$$

and the variance is calculated as

$$\text{Var}[x] = \sigma^2 = \beta^2 \left(\Gamma\left(1 + \frac{2}{\alpha}\right) - \Gamma\left(1 + \frac{1}{\alpha}\right)^2\right). \tag{6.65}$$

The binomial distribution is used to model the number of successes in a sample of size k drawn with replacement from a population of size N. This distribution was derived by Jacob Bernoulli (1655–1654), a prominent mathematician born in Basel, Switzerland. The probability density function is given by

$$p_X(x) = \sum_{k=0}^{\infty} \binom{N}{k} p^k (1-p)^{N-k} \delta(x-k).$$

Its characteristic function is written as

$$P_X(\omega) = [1 - p + pe^{-j\omega}]^N.$$

The Erlang distribution represents a sum of k independent exponential variables that have individual means $1/\lambda$. Agner Krarup Erlang (1878–1929) was a Danish mathematician, statistician and engineer, who made seminal contributions to the fields of traffic engineering and queuing theory.

The distribution also represents the time elapsed until the kth event of a Poisson process, with a rate of λ. It has the following probability density function

$$p_X(x) = \frac{\lambda^k x^{k-1} e^{-\lambda x}}{(k-1)!},$$

for $x, \lambda \geq 0$.

It has a characteristic function given by

$$P_X(\omega) = [\frac{\lambda}{\lambda + j\omega}]^N, \ \lambda > 0, \ N = 1, 2, \ldots.$$

The main moments of the distribution are: $E[X] = N/a$, $E[X^2] = N(N+1)/a^2$ and $\sigma_X^2 = N/a^2$.

The Poisson distribution expresses the probability of a given number of events that occur in a fixed interval of time or space, considering that the events happen with a constant mean rate and independently of the time since the last event. It is named after the French mathematician Siméon Denis Poisson (1781–1840), who made contributions to several fields, including statistics, complex analysis, partial differential equations, the calculus of variations, analytical mechanics, electricity and magnetism, thermodynamics, elasticity, and fluid mechanics.

The probability density function is

$$p_X(x) = e^{-b} \sum_{k=0}^{\infty} \frac{b^k}{k!} \delta(x-k),$$

and its cumulative distribution function is

$$P_X(x) = e^{-b} \sum_{k=0}^{\infty} \frac{b^k}{k!} u(x-k).$$

The characteristic function of the Poisson distribution is given by

$$P_X(\omega) = e^{-b(1-e^{j\omega})},$$

and the main moments of the distribution are: $E[X] = b$, $E[X^2] = b + b^2$.

The Maxwell distribution was named after James Clerk Maxwell (1831–1879), a Scottish mathematician and scientist who developed the classical theory of electromagnetic radiation, the first theory to describe electricity, magnetism and light as different manifestations of the same phenomenon. It is given by

$$p_Y(y) = \sqrt{\frac{2}{\pi}} \frac{y^2}{\sigma^3} e^{-\frac{y^2}{2\sigma^2}} u(y).$$

The main moments of the distribution are:

$$E[X] = 2\sigma \sqrt{\frac{2}{\pi}}$$

and

$$E[X^2] = 3\sigma^2.$$

6.9 Carl Friedrich Gauss

Johann Carl Friedrich Gauss (1777–1855), whose picture is shown in Figure 6.35, was a German mathematician and physicist who made important contributions to many fields in Mathematics and Physics. A mathematical prodigy, Gauss completed his main work *Disquisitiones Arithmeticae*, in 1798, when he was merely 21 years old. The paper, published three years later, in 1801, was fundamental to consolidate number theory as a discipline.

Gauss collaborated with the Wilhelm Eduard Weber (1804–1891), a German physics professor at the University of Göttingen, that led to representing the unit of magnetism in terms of mass, charge and time, and to the discovery of Kirchhoff's circuit laws in electricity. They also constructed the first electromechanical telegraph, in 1833, which connected the observatory with the Institute for Physics, in Göttingen.

Gustav Robert Kirchhoff (1824–1887) was a German physicist who contributed to the understanding of electrical circuits, spectroscopy, and the emission of black-body radiation by heated objects. Kirchhoff formulated

Figure 6.35 A portrait of Johann Carl Friedrich Gauss. Adapted from: Public Domain, https://commons.wikimedia.org/w/index.php?curid=6886354

his circuit laws, which are basic for Electrical Engineering, in 1845, while he was a student.

Gauss, Weber and Carl Wolfgang Benjamin Goldschmidt (1807–1851), a German astronomer, mathematician, and physicist, jointly published the *Atlas des Erdmagnetismus: nach den Elementen der Theorie entworfen* (Atlas of Geomagnetism: Designed According to the Elements of the Theory), a set of important geomagnetic maps.

6.10 Problems

1. Explain the following passage,
$$P_X(x) = P\{\omega \in \Omega : -\infty < X(\omega) \leq x\} = P\{X^{-1}((-\infty, x]) \in \mathcal{F}, \forall x \in \mathbb{R}\}.$$

2. Use set theoretic operations to demonstrate that the mathematical expectation has the following property,
$$\mathrm{E}[X + Y] = \mathrm{E}[X] + \mathrm{E}[Y].$$

3. A nonlinear system is described by the equation $Y = e^X$, $X \in \mathbb{R}$. Given the distribution $P_X(x)$ for the input variable, compute the distribution for the output variable? What is the pdf of the output random variable?

4. Compute the cumulative distribution function for the following probability density function,
$$p_Y(y) == \frac{1}{\pi\sqrt{V^2 - y^2}}.$$

5. The pdf for a random variable with Bernoulli distribution is given by
$$p_X(x) = p\,\delta(x + a) + q\,\delta(x - b).$$
Compute the expected value and the standard deviation of the random variable.

6. Compute the second moment for a random variable that has a geometric distribution, as follows,
$$p_X(x) = \sum_{k=0}^{\infty} (1 - \rho)\rho^k \delta(x - k).$$

7. A random variable has the following characteristic function. Calculate its second moment and its variance.
$$P_X(\omega) = \frac{\alpha}{(\alpha + j\omega)}.$$

8. The characteristic function for the binomial distribution is given by the following formula. Compute the first and second moments of the distribution.
$$P_X(\omega) = [1 - p + pe^{-j\omega}]^N.$$

7

Joint Random Variables

"We are usually convinced more easily by reasons we have found ourselves than by those which occurred to others."
Blaise Pascal

7.1 An Extension of the Concept of Random Variable

It is useful to extend the concept of random variables to address more complex and interesting problems. A joint random variable represents a mapping of joint random events, defined as a result of operations carried out in an algebra of sets. It is a transposition of events of the sample space Ω, to Borel rectangles, in the set of real numbers.

The way the mathematician Félix Edouard Juston Émile Borel structured the family of sets, or events, that bear his name, permits a simple generalization of the probability measure for any dimension.

The joint random variables form a natural extension of the concept of random variables. For two variables, the regions become rectangles in a wider sense. The following examples illustrate two-dimensional Borel sets, that is, for $\Omega = \mathbb{R}^2$.

Example: The rectangle $R = \{-\infty < X \leq x, -\infty < Y \leq y\}$ is illustrated in Figure 7.1. Calling this region R, the two-dimensional cumulative function can be defined as

$$P_{XY}(x, y) = P(R) = P\{-\infty < X \leq x, -\infty < Y \leq y\}. \blacksquare \quad (7.1)$$

The definition of random variable can be directly extended to a multidimensional space, as follows. The real functions $X_1, X_2, \ldots, X_n$ are random variables if and only if $\forall x_1, x_2, \ldots, x_n \in \mathbb{R}$, the following relations are satisfied $\{\omega : X_1(\omega) \leq x_1, X_2(\omega) \leq x_2, \ldots, X_n(\omega) \leq x_n\} \in \mathcal{F}$.

168 Joint Random Variables

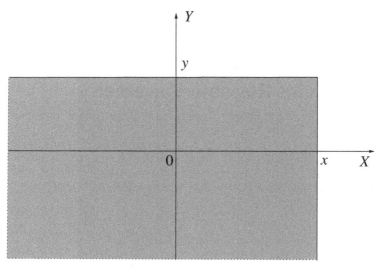

Figure 7.1 Region $\{-\infty < X \leq x,\ -\infty < Y \leq y\}$.

If f is an n-dimensional Borelian function defined in the considered space, and $X_1, X_2, \ldots, X_n$ are random variables, also called a vector, then the function $f(X_1, X_2, \ldots, X_n)$ is also a random variable.

To prove this statement, it is sufficient to consider the previous vector, defined as $X = X_1, X_2, \ldots, X_n$. Using the property of inverse functions, it can be shown that

$$f^{-1}(X)(\mathcal{B}(\mathbb{R})) = X^{-1}\left[f^{-1}(X)(\mathcal{B}(\mathbb{R}))\right] \subset X^{-1}(\mathcal{B}(\mathbb{R}^n)) \in \mathcal{F}. \quad (7.2)$$

The two-dimensional Cumulative Distribution Function (CDF) is the base for the definition of other mathematical constructions, as long as they are constructed from a measurable space $(\Omega, \mathcal{F}, \mathcal{P})$, in which the universe set is the real plane, $\Omega = \mathbb{R}^2$, $\mathcal{F} = \mathcal{B}(\mathbb{R}^2)$ represents the family of Borel rectangles and P is the measure of probability.

Example: The region $R_1 = \{-\infty < X \leq x_1,\ y_1 < Y \leq y_2\}$, shown in Figure 7.2, also defines a rectangle. It can be shown that the measure of that region can be expressed as a function of the joint CDF,

$$P\{-\infty < X \leq x_1,\ y_1 < Y \leq y_2\} = P_{XY}(x, y_2) - P_{XY}(x, y_1). \blacksquare \quad (7.3)$$

The following example shows how to obtain the pdf from the CDF in two dimensions, as long as infinitesimal rectangles, whose measures converge in the limit, are defined.

7.1 An Extension of the Concept of Random Variable

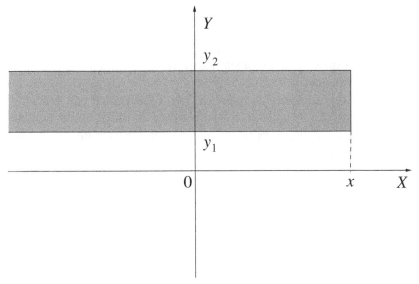

Figure 7.2 Region $\{-\infty < X \leq x_1,\ y_1 < Y \leq y_2\}$.

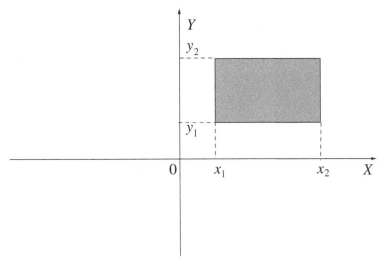

Figure 7.3 Region $\{x_1 < X \leq x_2,\ y_1 \leq Y \leq y_2\} = R_2$.

Example: The region $R_2 = \{x_1 < X \leq x_2,\ y_1 \leq Y \leq y_2\} = R_2$ has probability measure given by

170 Joint Random Variables

$$P(R_2) = P_{XY}(x_2, y_2) - P_{XY}(x_1, y_2) - P_{XY}(x_2, y_1) + P_{XY}(x_1, y_1). \blacksquare \tag{7.4}$$

Using the preceding expression, and following the same steps taken in the study of unidimensional variables, the relations between the joint pdf and CDF can be obtained.

Expression 7.4 is useful to compute the two-dimensional pdf. Defining $\Delta x = x_2 - x_1$ and $\Delta y = y_2 - y_1$, the pdf of the two-dimensional variable indicated in Figure 7.3 can be obtained, dividing the probability $P(R_2)$ by the area of the rectangle, that is,

$$\begin{aligned} p_{XY}(x, y) &= \frac{P_{XY}(x_1 + \Delta x, y_1 + \Delta y) - P_{XY}(x_1, y_2 + \Delta y)}{\Delta x \Delta y} \\ &\quad - \frac{P_{XY}(x_1 + \Delta x, y_1) - P_{XY}(x_1, y_1)}{\Delta x \Delta y}. \end{aligned} \tag{7.5}$$

Taking the limits $\Delta x \to 0$ and $\Delta y \to 0$, the two-dimensional pdf is obtained as

$$p_{XY}(x, y) = \frac{\partial^2 P_{XY}(x, y)}{\partial x \partial y}. \tag{7.6}$$

The CDF is naturally obtained with the Fundamental Theorem of Calculus, which results in

$$P_{XY}(x, y) = \int_{-\infty}^{x} \int_{-\infty}^{y} p_{XY}(\alpha, \beta) \, d\alpha \, d\beta. \tag{7.7}$$

Example: The Rician distribution is used to model communication channels that present multiple paths, with a line of sight between the transmitter and the receptor.

The joint distribution, for amplitude and phase, which is the origin of the Rician model for amplitude variations, is given by

$$p_{X\Theta}(x, \theta) = \frac{x e^{-A^2/2}}{2\pi \sigma^2} e^{-(x^2 - 2xA \cos \theta)/2\sigma^2}, \ 0 \leq x < \infty, \ -\pi \leq \theta \leq \pi. \tag{7.8}$$

Integrating the second-order cumulative function, in relation to the phase, θ, the marginal distribution, or first-order Rician pdf, is obtained,

7.2 Properties of Probability Distributions

$$p_X(x) = \frac{x}{\sigma^2} e^{-\frac{x^2+A^2}{2\sigma^2}} I_0(\frac{xA}{\sigma^2}) u(x),$$

in which $I_0(\cdot)$ is the modified first-order Bessel function, given by Formula 6.53.

Integrating the formula for all values of x yields the marginal distribution for the phase θ,

$$p_\Theta(\theta) = \frac{e^{-s^2}}{2\pi} + \frac{1}{2}\sqrt{\frac{s^2}{\pi}} \cos\theta \, e^{-s^2 \sin^2\theta}[1+2(1-Q(s/\sqrt{2}))\cos\theta]. \blacksquare \quad (7.9)$$

This expression generates a bell-shaped curve for large values of the signal-to-noise ratio (SNR). For $A = 0$ the distribution converges to uniform (Schwartz, 1970). The SNR is an auxiliary parameter given by $s^2 = A^2/2\sigma^2$. The function $Q(x)$ is defined in the usual form

$$Q(x) = \frac{1}{\sqrt{2\pi}} \int_x^\infty e^{-\frac{y^2}{2}} dy. \quad (7.10)$$

7.2 Properties of Probability Distributions

The joint probability density function and the joint cumulative probability function inherit the following basic properties from the definition of the Lebesgue measure, as in the following:

1. The cumulative function is non-negative – $P_{XY}(x,y) \geq 0$. This is a direct result of the Lebesgue measure definition.

2. The probability density function is also non-negative – $p_{XY}(x,y) \geq 0$.

3. The probability of an empty set is zero – $P_{XY}(-\infty,-\infty) = P(\emptyset) = 0$. This means that the empty set has measure zero.

4. The probability of the universal set is one – $P_{XY}(-\infty,\infty) = P(\Omega) = 1$. The universal set includes all the relevant elements for the problem of interest.

5. The marginal density in y is calculated integrating the joint distribution for x, that is,

$$p_Y(y) = \int_{-\infty}^\infty p_{XY}(x,y) \, dx. \quad (7.11)$$

6. The marginal density in x is calculated integrating the joint distribution for y, that is,

$$p_X(x) = \int_{-\infty}^{\infty} p_{XY}(x,y)\,dy. \quad (7.12)$$

Example: The Gaussian two-dimensional pdf is given by the formula

$$p_{XY}(x,y) = \frac{1}{2\pi\sigma_X\sigma_Y\sqrt{1-\rho^2}} e^{-\frac{1}{2(1-\rho^2)}\left[\frac{(x-m_X)^2}{\sigma_X^2} - \frac{2\rho(x-m_X)(y-m_Y)}{\sigma_X\sigma_Y} + \frac{(y-m_Y)^2}{\sigma_y^2}\right]}, \quad (7.13)$$

in which ρ represents the correlation coefficient, m_X and m_Y are the respective means for the random variables X and Y, σ_X and σ_Y are the respective standard deviations. ∎

7.3 Moments in Two Dimensions

Consider that the random variables X and Y, are defined in the probability space $(\mathbb{R}, \mathcal{B}(\mathbb{R}), P)$. In general, the statistical expectancy of a given function $f(x,y)$, of the variables X and Y, is defined by

$$E[f(x,y)] = \int_{-\infty}^{\infty}\int_{-\infty}^{\infty} f(x,y)p_{XY}(x,y)\,dx\,dy. \quad (7.14)$$

For the random variables X and Y their joint moments of order $n+m$ can be defined with the use of the Riemann integral, considering that $f(x,y) = x^n y^m$.

$$E[X^n Y^m] = \int_{-\infty}^{\infty}\int_{-\infty}^{\infty} x^n y^m p_{XY}(x,y)\,dx\,dy, \quad \text{order}: n+m. \quad (7.15)$$

For the random variables X and Y, defined in the probability space of real numbers, the covariance, which is a measure of correlation between variables, is defined, for $f(x,y) = (x-m_X)(y-m_Y)$, as

$$C[X,Y] = E[(X-m_X)(Y-m_Y)] = E[XY] - E[X]\cdot E[Y]. \quad (7.16)$$

When there is no correlation between the variables, then $C[X,Y] = 0$

7.3 Moments in Two Dimensions

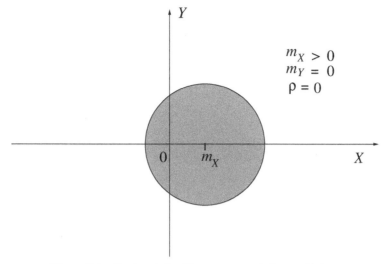

Figure 7.4 Region defined by a zero correlation coefficient.

The correlation coefficient, ρ_{XY}, or simply ρ, is the covariance normalized by the product of the standard deviations

$$\rho_{XY} = \frac{C[X,Y]}{\sigma_X \sigma_Y}. \tag{7.17}$$

The correlation coefficient is limited to the interval $[-1, 1]$, that is, $-1 \leq \rho_{XY} \leq 1$, and the variables are said to be uncorrelated when $\rho_{XY} = 0$. When $\rho_{XY} \approx 1$, it is said that the variables are positively and strongly correlated, In the case in which $\rho_{XY} \approx -1$, the variables are negatively correlated.

Figures 7.4, 7.5, and 7.6 illustrate the regions defined by different correlation coefficients and expected values. The standard deviations of X and Y also distort the shape of the regions.

The correlation between X and Y is a measure of the orthogonality between the variables, which is sometimes confusing. The correlation is given by the expected mean of the product between the random variables

$$R[X, Y] = E[XY]. \tag{7.18}$$

The correlation can be written as

$$R[X, Y] = C[X, Y] + E[X] \cdot E[Y], \tag{7.19}$$

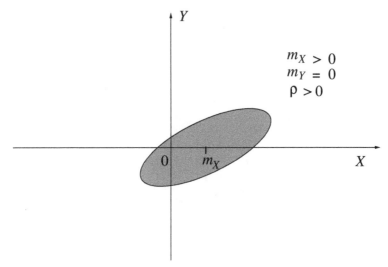

Figure 7.5 Region defined by a positive correlation coefficient.

and, for two random variables to be uncorrelated, the correlation must be equal to the product of the means, that is,

$$R[X, Y] = E[X] \cdot E[Y] = m_X \cdot m_Y.$$

The correlation has some useful properties that help establish limits for calculations. For example, it can be related to the sum of the second moments of X and Y,

$$R[X, Y] \leq \frac{E[X^2] + E[Y^2]}{2}, \qquad (7.20)$$

or can be related to the product of the second moments of X and Y,

$$R[X, Y] \leq \sqrt{E[X^2] \cdot E[Y^2]}. \qquad (7.21)$$

To demonstrate the second property, it suffices to use the following tautology, which is always true,

$$E[(\alpha X - Y)^2] \geq 0.$$

Calculating the square of the binomial yields

$$E[\alpha^2 X^2 - 2\alpha XY + Y^2] \geq 0.$$

Using the linearity of the expected value,

$$\alpha^2 E[X^2] - 2\alpha E[XY] + E[Y^2] \geq 0,$$

7.3 Moments in Two Dimensions

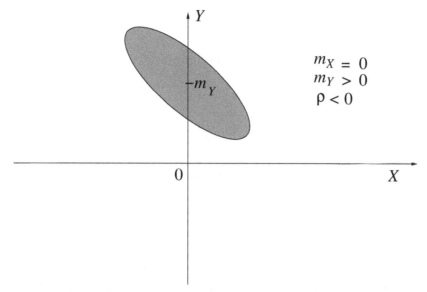

Figure 7.6 Region defined by a negative correlation coefficient.

that results in
$$\alpha^2 E[X^2] - 2\alpha R[X,Y] + E[Y^2] \geq 0,$$
which is an inequality in terms of the variable α, while the expected values act as parameters for the inequality.

For the parabola to be above, or at least, touching the X-axis, the discriminant of the equation must be negative, that is,
$$\Delta = 4R^2[X,Y] - 4E[X^2] \cdot E[Y^2] \leq 0.$$

Therefore, after simplifying, one obtains
$$R^2[X,Y] \leq E[X^2] \cdot E[Y^2],$$
and the property follows. To demonstrate the first property, is suffices to attribute $\alpha = 1$ and repeat the deduction.

Example: It is important to note that the lack of correlation does not imply that the variables are independent, which is a more strict condition, and usually difficult to prove. Suppose that
$$X = V \cos \theta$$

and
$$Y = V \sin \theta,$$

in which θ is a random variable with uniform distribution in the interval $[0, 2\pi]$, and V is a constant.

The means of X and Y are zero, because

$$\begin{aligned} \mathrm{E}[X] = \mathrm{E}[V \cos \theta] &= \int_{-\infty}^{\infty} V \cos \theta \frac{1}{2\pi} [\mathrm{u}(\theta + \pi) - \mathrm{u}(\theta - \pi)] d\theta \\ &= \frac{V}{2\pi} \int_0^{2\pi} \cos \theta d\theta = \frac{V}{2\pi} [\sin \theta]_0^{2\pi} = 0. \end{aligned} \quad (7.22)$$

and

$$\mathrm{E}[Y] = \mathrm{E}[V \sin \theta] = \frac{V}{2\pi} \int_0^{2\pi} \sin \theta d\theta = \frac{V}{2\pi} [\cos \theta]_0^{2\pi} = 0.$$

The correlation between the variables is zero, which is shown as follows

$$\mathrm{E}[XY] = V^2 \mathrm{E}[\cos \theta \cdot \sin \theta] = \frac{V^2}{2\pi} \int_0^{2\pi} \cos \theta \sin \theta d\theta = 0,$$

because the sine and cosine are orthogonal functions.

However, the sum of squares of X and Y results in $X^2 + Y^2 = V^2$, indicating that the variables are dependent. ∎

The variance of the sum of the uncorrelated variables X and Y is given by

$$\begin{aligned} \mathrm{V}[X + Y] &= \mathrm{E}\left[((X + Y) - \mathrm{E}[X + Y])^2\right] \\ &= \mathrm{E}\left[(X + Y)^2\right] - (\mathrm{E}[X] + \mathrm{E}[Y])^2 \\ &= \mathrm{E}\left[X^2\right] - \mathrm{E}^2[X] + \mathrm{E}\left[Y^2\right] - \mathrm{E}^2[Y] \\ &\quad + 2\left(\mathrm{E}[XY] - \mathrm{E}[X] \cdot \mathrm{E}[Y]\right) \\ &= \mathrm{V}[X] + \mathrm{V}[Y], \end{aligned} \quad (7.23)$$

because $\mathrm{E}[XY] = \mathrm{E}[X] \cdot \mathrm{E}[Y]$, for uncorrelated variables.

Example: If the correlation coefficient is null, $\rho = 0$, the joint probability density function simplifies to

$$p_{XY}(x, y) = \frac{1}{2\pi \sigma_X \sigma_Y} e^{-\left[\frac{(x - m_X)^2}{2\sigma_X^2} + \frac{(y - m_Y)^2}{2\sigma_Y^2}\right]},$$

which can be written as the product of the marginal distributions,

$$p_{XY}(x,y) = \frac{1}{\sigma_X\sqrt{2\pi}} e^{-\frac{(x-m_X)^2}{2\sigma_X^2}} \cdot \frac{1}{\sigma_Y\sqrt{2\pi}} e^{-\frac{(y-m_y)^2}{2\sigma_y^2}} = p_X(x) \cdot p_Y(y),$$

that is, decorrelation implies independence for Gaussian random variables.
∎

7.4 Conditional Moments

For the random variables X and Y, defined in the probability space $(\mathbb{R}, \mathcal{B}((\mathbb{R}), P)$, the conditional expectancy of X given $Y = y$, is defined using the Stieltjes integral, as

$$\mathrm{E}[X|Y = y] = \int_{-\infty}^{\infty} x \, dP_{X|Y}(x|Y = y), \qquad (7.24)$$

in which $P_{X|Y}(x|Y = y)$ is the conditional probability distribution of X given $Y = y$.

One must remember that $\mathrm{E}[X|Y = y]$ is a function of y, that can be written as $\mathrm{E}[X|y]$, known in Statistics as the regression of X in relation to Y. It represents the expectancy of X given the information about the occurrence of Y.

This concept is important for the definition of conditional entropy, mutual information and channel capacity. In case the value of Y is not specified, then $\mathrm{E}[X|Y]$ is a function of the random variable Y, for which the expected value is $\mathrm{E}[\mathrm{E}[X|Y]] = \mathrm{E}[X]$, demonstrated as follows.

Because the variables are continuous, which allows the use of the Riemann integral, and the joint pdf is known, then

$$\begin{aligned}
\mathrm{E}[X|Y = y]] &= \int_{-\infty}^{\infty} x \, dP_{X|Y}(x|Y = y) \\
&= \int_{-\infty}^{\infty} x \, p_{X|Y}(x|y) dx \\
&= \int_{-\infty}^{\infty} x \frac{p_{XY}(x,y)}{p_Y(y)} dx, \; p_Y(y) > 0. \qquad (7.25)
\end{aligned}$$

Therefore,

$$\mathrm{E}[\mathrm{E}[X|Y]] = \int_{-\infty}^{\infty} \mathrm{E}[X|Y = y] dP_Y(y)$$

$$= \int_{-\infty}^{\infty} \left(\int_{-\infty}^{\infty} x \frac{p_{XY}(x,y)}{p_Y(y)} dx \right) p_Y(y) dy$$

$$= \int_{-\infty}^{\infty} x \left(\int_{-\infty}^{\infty} p_{XY}(x,y) dy \right) dx$$

$$= \int_{-\infty}^{\infty} x p_X(x) dx$$

$$= \mathrm{E}[X]. \tag{7.26}$$

Example: The expected value of the product $X \cdot Y$, in terms of the conditional expectancy of X, given that Y has occurred, can be calculated from the previous result.

$$\mathrm{E}[XY|Y=y] = \mathrm{E}[Xy|Y=y] = y\mathrm{E}[X|Y=y],$$

therefore,

$$\mathrm{E}[XY|Y] = Y\mathrm{E}[X|Y].$$

Calculating the expected value yields

$$\mathrm{E}[XY] = \mathrm{E}[Y\mathrm{E}[X|Y]].\blacksquare$$

For the random variables X and Y, defined in the probability space $(\mathbb{R}, \mathcal{B}(\mathbb{R}), P)$, the conditional variance is given by

$$\begin{aligned} \mathrm{V}[X|Y=y] &= \mathrm{E}[(X - \mathrm{E}[X|Y=y])^2] \\ &= \mathrm{E}[X^2|Y=y] - \mathrm{E}^2[X|Y=y], \end{aligned} \tag{7.27}$$

which represents the expected value of the deviations of the variable in relation to the conditional expectancy.

For the random variables X, Y and Z, defined in the probability space $(\mathbb{R}, \mathcal{B}(\mathbb{R}), P)$, the conditional covariance is given by

$$\mathrm{C}[X,Y|Z=z] = \mathrm{E}[XY|Z=z] - \mathrm{E}[X|Z=z] \cdot \mathrm{E}[Y|Z=z], \tag{7.28}$$

considering the existence of the statistical expectancies.

7.5 Two-Dimensional Characteristic Function

The two-dimensional characteristic function is defined as the two-dimensional Fourier transform of the joint probability density function $p_{XY}(x,y)$. It is particularly appropriate for the calculation of moments and

7.5 Two-Dimensional Characteristic Function

the sum of random variables, and is commonly used to demonstrate the Central Limit Theorem.

The definition of the characteristic function in two dimensions is the generalization of the unidimensional function

$$P_{XY}(\omega, \nu) = \mathrm{E}[e^{-j\omega X - j\nu Y}], \tag{7.29}$$

that is,

$$P_{XY}(\omega, \nu) = \int_{-\infty}^{\infty} \int_{-\infty}^{\infty} p_{XY}(x, y) e^{-j\omega x - j\nu y} \, dx \, dy. \tag{7.30}$$

Therefore, the pdf can be obtained with the inverse transform

$$p_{XY}(x, y) = \frac{1}{4\pi^2} \int_{-\infty}^{\infty} \int_{-\infty}^{\infty} P_{XY}(\omega, \nu) e^{j\omega x + j\nu y} \, d\omega \, d\nu. \tag{7.31}$$

If the random variables are independent,

$$P_{XY}(\omega, \nu) = \mathrm{E}[e^{-j\omega X}] \cdot \mathrm{E}[e^{-j\nu Y}] = P_X(\omega) \cdot P_Y(\nu). \tag{7.32}$$

For a sum of N independent random variables $Y = X_1 + X_2 + \cdots + X_N$, one may write

$$P_Y(\omega) = \prod_{k=1}^{N} P_{X_k}(\omega). \tag{7.33}$$

And, in case they are identically distributed, then

$$P_Y(\omega) = P_X^N(\omega). \tag{7.34}$$

Applying the inverse Fourier transform, results in a pdf that can be written as the convolution of the original functions

$$p_Y(y) = \underbrace{p_X(x) * \cdots * p_X(x)}_{N \text{ times}}. \tag{7.35}$$

Example: An important application for the property of the sum of independent random variables is the power-flow in power systems analysis. Usually, the companies determine maximum, mean, and minimum values for residential consumption, and later compute the power-flow in the network for the three cases, and this has an elevated computational cost.

However, the energy consumption is evidently random, and it does not make sense to consider that all users have the same electricity usage.

180 Joint Random Variables

Suppose that the energy consumption for each house, X_k, is characterized by a probability distribution $p_{X_k}(x)$, with mean value m_{X_k} and standard deviation σ_{X_k}.

The total load-flow of a city with N houses is modeled by the distribution of the sum of all the individual load-flows, $Y = X_1 + X_2 + \cdots + X_N$, that is,

$$P_Y(\omega) = \prod_{k=1}^{N} P_{X_k}(\omega),$$

in which $P_{X_k}(\omega)$ represents the characteristic function of each consumer. The calculation can be quickly done using the Fast Fourier Transform (FFT) algorithm. ∎

After obtaining the characteristic function of the total load-flow, it is sufficient to calculate the Inverse Fourier Transform (IFT), using the same algorithm, to obtain the pdf that represents the energy consumption for that network. From this distribution, it is possible to calculate all the moments, including the mean and standard deviation, to obtain a statistical estimate of the total consumption.

The bi-dimensional characteristic function, also called moment generating function, can be used to calculate the mean, standard deviation, and all the moments of the random variables X and Y,

$$\mathrm{E}[X^n Y^m] = \frac{1}{(-j)^{n+m}} \cdot \frac{\partial^{n+m}}{\partial \omega^n \partial \nu^m} P_{XY}(\omega, \nu) \bigg|_{\omega=0, \nu=0}. \qquad (7.36)$$

7.5.1 Sum of Random Variables

The characteristic function is also used to determine the pdf of a random variable that is the sum of other independent random variables, $Z = X + Y$.

For this particular case, the characteristic function of the resulting random variable is

$$P_Z(\omega) = \mathrm{E}[e^{-j\omega Z}] = \mathrm{E}[e^{-j\omega(X+Y)}] = P_X(\omega) \cdot P_Y(\omega). \qquad (7.37)$$

Using a property of Fourier transform, it is possible to write,

$$p_Z(z) = \int_{-\infty}^{\infty} p_X(\rho) p_Y(z - \rho) d\rho, \qquad (7.38)$$

or, in an equivalent form,

$$p_Z(z) = \int_{-\infty}^{\infty} p_X(z-\rho)p_Y(\rho)d\rho. \tag{7.39}$$

Therefore, the sum of independent random variables results in the convolution of their respective probability density functions. This is a powerful property, usually stated as a theorem, that can be used to prove the Central Limit Theorem, and also to determine the bit, or symbol, error probability for a channel with additive noise.

7.6 Function of Joint Random Variables

The transformation of joint random variables, also called, function of joint random variables, can be used to solve problems in several areas of Mathematics, Physics, Engineering, Economics, to mention only a few. It is also useful in Electronics, Communications, Control and Power Systems.

In order to obtain the formula that maps the input random variables (X, Y) into the output random variables (U, V), it is necessary to take into account that the probability measure is preserved for any transformation, as depicted in Figure 7.7.

That is, the differential volumes at the input and output are equal

$$p_{XY}(x, y)dxdy = p_{UV}(u, v)dudv, \tag{7.40}$$

in which $dE = dxdy$ and $dS = dudv$ represent, respectively, the differential areas that are spanned by the joint input and output variables.

In Figure 7.8 the square represents the differential area produced by the input random variables. The vertices of the Borel rectangle are the

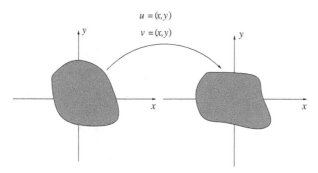

Figure 7.7 Generic regions used to analyze the transformation.

coordinates $A = x, y + dy$, $B = x + dx, y + dy$, $C = x + dx, y$ and $D = x, y + dy$.

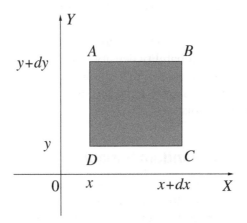

Figure 7.8 Region $\{x < X \leq x + dx,\ y < Y \leq y + dy\}$.

At the output, the joint random variable is transformed by the respective functions, $U = f(X, Y)$ and $V = g(X, Y)$. This mapping produces a new differential area, which is shown in Figure 7.9, whose vertices are

$$E = (f(x, y + dy), g(x, y + dy)),$$
$$F = (f(x + dx, y + dy), g(x + dx, y + dy)),$$
$$G = (f(x + dx, y), g(x + dx, y)),$$
$$H = (f(x, y), g(x, y)).$$

Because the area is infinitesimal, it can be approximated by the parallelogram illustrated in Figure 7.10, and the vertices, obtained by the linearization process, are

$$E = \left(u + \frac{\partial f(x, y)}{\partial y} dy,\ v + \frac{\partial g(x, y)}{\partial y} dy\right),$$

$$F = \left(u + \frac{\partial f(x, y)}{\partial x} dx + \frac{\partial f(x, y)}{\partial y} dy,\ v + \frac{\partial g(x, y)}{\partial x} dx, + \frac{\partial g(x, y)}{\partial y} dy\right),$$

$$G = \left(u + \frac{\partial f(x, y)}{\partial x} dx,\ v + \frac{\partial g(x, y)}{\partial x} dx\right),$$

$$H = (u, v).$$

7.6 Function of Joint Random Variables

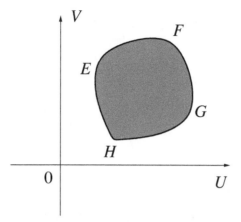

Figure 7.9 Differential region for the output joint variables.

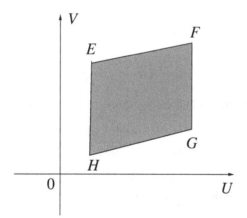

Figure 7.10 Linearized output differential region.

The area of the parallelogram, with vertices $A = (x_4, y_4)$, $B = (x_3, y_3)$, $C = (x_2, y_2)$, $D = (x_1, y_1)$, is given by the formula (Bronstein and Semendiaev, 1979)

$$S = (x_1 - x_2)(y_1 + y_2) + (x_2 - x_3)(y_2 + y_3) + (x_3 - x_1)(y_3 + y_1). \quad (7.41)$$

Therefore, one obtains

$$dS = \left(u - u - \frac{\partial f(x,y)}{\partial x}dx\right)\left(v + v + \frac{\partial g(x,y)}{\partial x}dx\right)$$

$$+ \left(u + \frac{\partial f(x,y)}{\partial x} dx - u - \frac{\partial f(x,y)}{\partial x} dx - \frac{\partial f(x,y)}{\partial y} dy \right)$$
$$\cdot \left(v + \frac{\partial g(x,y)}{\partial x} dx + v + \frac{\partial g(x,y)}{\partial y} dy \right)$$
$$+ \left(u - u + \frac{\partial f(x,y)}{\partial x} dx + \frac{\partial f(x,y)}{\partial y} dy \right)$$
$$\cdot \left(v + v + \frac{\partial g(x,y)}{\partial x} dx + \frac{\partial g(x,y)}{\partial y} dy \right).$$

This result can be simplified to

$$dS = \left(-\frac{\partial f(x,y)}{\partial x} dx \right) \left(2v + \frac{\partial g(x,y)}{\partial x} dx \right)$$
$$+ \left(-\frac{\partial f(x,y)}{\partial y} dy \right) \left(2v + \frac{\partial g(x,y)}{\partial x} dx + \frac{\partial g(x,y)}{\partial y} dy \right)$$
$$+ \left(\frac{\partial f(x,y)}{\partial x} dx + \frac{\partial f(x,y)}{\partial y} dy \right) \left(2v + \frac{\partial g(x,y)}{\partial x} dx + \frac{\partial g(x,y)}{\partial y} dy \right).$$

Multiplying the terms and simplifying the result, gives

$$dS = \left| \frac{\partial f(x,y)}{\partial x} \frac{\partial g(x,y)}{\partial y} - \frac{\partial g(x,y)}{\partial x} \frac{\partial f(x,y)}{\partial y} \right| dxdy.$$

The last expression represents the Jacobian of the transformation, and can be written using the determinant notation

$$J(x,y) = \begin{vmatrix} \partial u/\partial x & \partial u/\partial y \\ \partial v/\partial x & \partial v/\partial y \end{vmatrix}. \tag{7.42}$$

The Jacobian is named after Carl Gustav Jacob Jacobi (1804–1851), a German mathematician who made fundamental contributions to elliptic functions, dynamics, differential equations, determinants, and number theory. Jacobi, born in Potsdam, Kingdom of Prussia, was the first Jewish mathematician appointed professor at a German university.

Substituting the result into Equation (7.40), yields

$$p_{UV}(u,v) = \frac{p_{XY}(x,y)}{|J(x,y)|}, \tag{7.43}$$

which is a compact formula to compute the pdf of the output joint random variables, given the input and the Jacobian.

7.6 Function of Joint Random Variables

Example: Consider the transformation obtained by the equations $U = aX + bY$ and $V = cX + dY$.

The Jacobian is

$$J(x, y) = \begin{vmatrix} a & b \\ c & d \end{vmatrix} = ad - cb$$

and the inverse functions are given by

$$x = \frac{1}{a}(u - by),$$

$$v = \frac{c}{a}(u - by) + dy \qquad \frac{a}{c}v = u - by + \frac{da}{c}y,$$

$$y\left(\frac{da}{c} - b\right) = \frac{a}{c}v - u \Rightarrow y = \left(\frac{a}{c}v - u\right)\frac{c}{da - cb},$$

$$y = \frac{uc - av}{cb - da},$$

$$x = \frac{1}{a}\left[u - \left(\frac{uc - av}{cb - da}\right)\right] = \frac{du - bv}{ad - bc}.$$

Therefore, the joint pdf at the output is given by

$$p_{UV}(u, v) = \frac{p_{XY}\left(\frac{du-bv}{ad-bc}; \frac{uc-av}{cb-bc}\right)}{|ad - cb|}. \blacksquare$$

Example: It is possible to solve $U = X + Y$ using an auxiliary variable $V = X$.

Using the previous formula,

$$p_{UV}(u, v) = \frac{p_{XY}(x, y)}{|J(x, y)|},$$

in which

$$p_U(u) = \int_{-\infty}^{\infty} p_{UV}(u, v) \, dv.$$

But,

$$J(x, y) = \begin{vmatrix} 1 & 1 \\ 1 & 0 \end{vmatrix} = -1,$$

then

$$p_{UV}(u, v) = \frac{p_{XY}(x, y)}{1} \qquad \text{in which}: x = v, \ y = u - v.$$

Therefore,
$$p_{UV}(u,v) = p_{XY}(v, u-v)$$
and
$$p_U(u) = \int_{-\infty}^{\infty} p_{XY}(v, u-v)\, dv. \blacksquare$$

If X and Y are independent random variables, the previous equation simplifies to

$$p_{XY}(x,y) = p_X(x) \cdot p_Y(y) \quad \Rightarrow \quad p_U(u) = \int_{-\infty}^{\infty} p_X(v) \cdot p_Y(u-v)\, dv.$$

This result indicates that the sum of random variables produces the convolution of the respective probability density functions, as previously established.

7.7 Transformation of Random Vectors

Consider that $f(\mathbf{x})$ is a vector-valued function of $\mathbf{x} \in \mathbb{R}^n$, and $\mathbf{X}$ is a random vector in the same space. Define the random vector $\mathbf{Y} = f(\mathbf{X})$, obtained from the input vector by the application of the function (Gubner, 2006).

$$\begin{bmatrix} y_1 \\ y_2 \\ \vdots \\ y_n \end{bmatrix} = \begin{bmatrix} f_1(x_1, x_2, \ldots, x_n) \\ f_2(x_1, x_2, \ldots, x_n) \\ \vdots \\ f_n(x_1, x_2, \ldots, x_n) \end{bmatrix}. \quad (7.44)$$

If the function f is invertible, it is possible to apply f^{-1} to both sides of the equation $\mathbf{y} = f(\mathbf{x})$ to obtain $f^{-1}(\mathbf{y}) = \mathbf{x}$. Using the notation $\mathbf{x} = g(\mathbf{y})$, it is possible to write the vector equation as

$$\begin{bmatrix} x_1 \\ x_2 \\ \vdots \\ x_n \end{bmatrix} = \begin{bmatrix} g_1(y_1, y_2, \ldots, y_n) \\ g_2(y_1, y_2, \ldots, y_n) \\ \vdots \\ g_n(y_1, y_2, \ldots, y_n) \end{bmatrix}. \quad (7.45)$$

Assuming that g is a continuous function, with continuous partial derivatives, define

$$\mathrm{d}g(\mathbf{y}) = \begin{bmatrix} \frac{\partial g_1}{\partial y_1} & \frac{\partial g_1}{\partial y_2} & \cdots & \frac{\partial g_1}{\partial y_n} \\ \frac{\partial g_2}{\partial y_1} & \frac{\partial g_2}{\partial y_2} & \cdots & \frac{\partial g_2}{\partial y_n} \\ & \vdots & & \\ \frac{\partial g_n}{\partial y_1} & \frac{\partial g_n}{\partial y_2} & \cdots & \frac{\partial g_n}{\partial y_n} \end{bmatrix}. \tag{7.46}$$

Next, it is necessary to compute the probability that the random vector $\mathbf{Y}$ is inside a certain region $\mathbf{A}$, this is, $P(\mathbf{Y} \in \mathbf{A}) = P(f(\mathbf{X}) \in \mathbf{A})$. One defines $\mathbf{B} = \{\mathbf{x} : f(\mathbf{x}) \in \mathbf{A}\}$, and obtains

$$P(\mathbf{Y} \in \mathbf{A}) = P(f(\mathbf{X}) \in \mathbf{A}) = P(\mathbf{X} \in \mathbf{B}), \tag{7.47}$$

which can be written as

$$P(\mathbf{Y} \in \mathbf{A}) = \int_{\mathbb{R}^n} I_{\mathbf{B}}(\mathbf{x}) p_{\mathbf{X}}(\mathbf{x}) \mathrm{d}\mathbf{x}, \tag{7.48}$$

in which $I_{\mathbf{B}}(\cdot)$ is the indicator function.

Applying the multivariate variable change $\mathbf{x} = g(\mathbf{y})$, considering that

$$\mathrm{d}\mathbf{x} = |\det \mathrm{d}g(\mathbf{y})| \mathrm{d}\mathbf{y}, \tag{7.49}$$

one obtains,

$$P(\mathbf{Y} \in \mathbf{A}) = \int_{\mathbb{R}^n} I_{\mathbf{B}}(g(\mathbf{y})) p_{\mathbf{X}}(g(\mathbf{y})) |\det \mathrm{d}g(\mathbf{y})| \mathrm{d}\mathbf{y}. \tag{7.50}$$

Consider that $I_{\mathbf{B}}(g(\mathbf{y})) = 1$ if and only if $g(\mathbf{y}) \in \mathbf{B}$, which occurs if and only if $f(g(\mathbf{y})) \in \mathbf{A}$. But, since g is the inverse function of f, $f(g(\mathbf{y})) = \mathbf{y}$, and $I_{\mathbf{B}}(g(\mathbf{y})) = I_{\mathbf{A}}(\mathbf{y})$. Therefore,

$$P(\mathbf{Y} \in \mathbf{A}) = \int_{\mathbf{A}} p_{\mathbf{X}}(g(\mathbf{y})) |\det \mathrm{d}g(\mathbf{y})| \mathrm{d}\mathbf{y}. \tag{7.51}$$

Because the original set $\mathbf{A}$ is arbitrary, the integrand must be the probability density function of $\mathbf{Y}$.

$$p_{\mathbf{Y}}(\mathbf{y}) = p_{\mathbf{X}}(g(\mathbf{y})) |\det \mathrm{d}g(\mathbf{y})|. \tag{7.52}$$

The preceding equations, which provide the multivariate generalization of Equation (7.42), are called Jacobian formulas.

7.8 Complex Random Variables

Sometimes it is useful to define a complex random variable, in order to deal with certain problems in Electrical Engineering, Physics or Mathematics. Let $\mathbb{R}$, as usual, denote the set of real numbers. The set of complex numbers $\mathbb{C}$ is the plane $\mathbb{R}^2$ equipped with complex addition and complex multiplication (Ericsson et al., 2009).

It is a pair of real random variables written in the form (Gubner, 2006),

$$Z = X + jY, \qquad (7.53)$$

in which j denotes $\sqrt{-1}$, the imaginary unit.

Geometrically a complex random variable may be interpreted as a random point Z in the complex plane, as shown in Figure 7.11 (Ventsel, 1973).

The generalization of the fundamental characteristics of the real random variables to the complex ones is not difficult. Those generalizations must be such that, if $Y = 0$, and Z is real, they reduce to the usual characteristics of the random variables.

The expected value of the complex random variable is given by

$$\mathrm{E}[Z] = \mathrm{E}[X] + j\mathrm{E}[Y] = m_X + jm_Y = m_Z. \qquad (7.54)$$

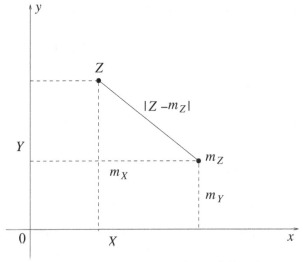

Figure 7.11 Representation of a complex random variable in the complex plane.

The expected value is a point around which there is a dispersion of the complex random variable Z.

The variance of Z is

$$V[Z] = E[(Z - E[Z])(Z - E[Z])^*] = E[|Z - E[Z]|^2] = V[X] + V[Y], \qquad (7.55)$$

in which the symbol $*$ denotes complex conjugate. That is, the variance is the expected value of the square of the modulus of the complex random variable.

On the other hand,

$$E[(Z - E[Z])^2] = V[X] - V[Y] + j2C[X, Y]. \qquad (7.56)$$

in which $C[X, Y]$ represents the covariance of X and Y. The expression is zero if and only if X and Y are uncorrelated and have the same variance.

The complex covariance is defined as

$$\begin{aligned} K[Z, R] &= E[(Z - E[Z])(R - E[R])^*] \\ &= C[X, U] + C[Y, V] + j\left(C[Y, U] - C[X, V]\right). \end{aligned} \qquad (7.57)$$

in which $R = U + jV$, and V, U are real variables.

The distribution of a complex random variable Z can be associated with the joint, or real bivariate, distribution of X and Y. The cumulative distribution function of the complex random variable Z is

$$P_Z(z) = P\{X \leq x, Y \leq y\} = P_{(X,Y)}(x, y), \qquad (7.58)$$

in which $(X, Y) = X + jY$, is another representation for the complex random variable, and $P_{(X,Y)}(x, y)$ is the joint cumulative distribution function of the random vector (X, Y).

The probability density function, if it exists, is defined as

$$p_Z(z) = p_Z(x + jy) = p_{(X,Y)}(x, y), \qquad (7.59)$$

such that

$$P_{(X,Y)}(x, y) = \int_{-\infty}^{\infty} \int_{-\infty}^{\infty} p_{(X,Y)}(x, y) dx dy, \qquad (7.60)$$

for all $(x, y) \in \mathbb{R}^2$. This implies that

$$\frac{\partial^2 P_{(X,Y)}(x, y)}{\partial x \partial y} = 2j \left(\frac{\partial^2 P_Z(z)}{\partial z^2} - \frac{\partial^2 P_Z(z)}{\partial z^{*2}} \right)$$

exists and equals $p_Z(z)$ everywhere.

Example: Sometimes the formula for the pdf can be expressed in a compact form in terms of the complex random variable Z. If X and Y are independent Gaussian variables, with zero mean and unit variance, $N(0,1)$, the formula for the joint pdf can be written as

$$p_{(X,Y)}(x,y) = \frac{1}{\sqrt{2\pi}}e^{-x^2/2} \cdot \frac{1}{\sqrt{2\pi}}e^{-y^2/2} = \frac{1}{2\pi}e^{-|z|^2/2}. \blacksquare$$

In the previous formula, $\mathrm{E}[Z] = 0$ and $\mathrm{V}[Z] = 2$. Also, the probability density function is circularly symmetric since $|z|^2 = x^2 + y^2$ depends only on the distance of point $(x,y) \in \mathbb{R}^2$ to the origin.

7.9 Félix Borel

Félix Edouard Juston Émile Borel (1871–1956) was a French mathematician and also a politician. He is known for his fundamental work in the areas of measure theory and probability. Figure 7.12 shows a portrait of Borel. His thesis at the *École Polytechnique*, published in 1893, was titled *Sur Quelques Points de la Théorie des Fonctions* (On Some Points in the Theory of Functions).

Figure 7.12 A portrait of Félix Edouard Juston Émile Borel. Adapted from: Public Domain,

Félix Émile Borel, René-Louis Baire, Henri Lebesgue, and Andrei Kolmogorov were the pioneers of measure theory and its application to probability theory. Along with his career as a mathematician, Borel was also active in politics. He was a member of the Chamber of Deputies and Minister of the Navy, in the cabinet of fellow mathematician Paul Painlevé.

Borel founded the *Institut de Statistiques de l'Université de Paris* (Paris Institute of Statistics), in 1922, the oldest and prestigious French graduate school for statistics, and co-founded the Institut Henri Poincaré in Paris, in 1928. The concept of a Borel set, widely used in probability theory, is named in his honor.

7.10 Problems

1. For the Gaussian two-dimensional pdf, given by the following formula, compute the marginal distribution of the random variable X.

$$p_{XY}(x,y) = \frac{1}{2\pi\sigma_X\sigma_Y\sqrt{1-\rho^2}} e^{-\frac{1}{2(1-\rho^2)}\left[\frac{(x-m_X)^2}{\sigma_X^2} - \frac{2\rho(x-m_X)(y-m_Y)}{\sigma_X\sigma_Y} + \frac{(y-m_y)^2}{\sigma_y^2}\right]},$$

2. Using the previous formula, prove that it is possible de separate the marginal distributions of X and Y if the correlation coefficient, ρ, is zero.

3. For the composite random variable $Z = X + Y$, considering that X independent of Y, prove that

$$p_Z(z) = \int_{-\infty}^{\infty} p_X(\rho) p_Y(z-\rho) \mathrm{d}\rho.$$

4. If X and Y are not independent in the previous question, determine a formula for the distribution of $Z = X - Y$.

5. Determine a formula for the distribution of $Z = \max\{X, Y\}$. Compute the probability density function of Z.

6. Let X and Y be independent univariate Gaussian random variables. Consider that two new variables are formed, $U = 2X - Y$ and $V = X - 2Y$, and compute the joint distribution of U and V. Determine if the random variables U and V are independent.

8
Probability Fundamental Inequalities

"If your experiment needs statistics, you ought to have done a better experiment."
Ernest Rutherford

8.1 Historical Notes

The Probability Theory took some time to establish itself as a branch of Mathematics because, in the beginning, it was associated with gambling. Pierre-Simon de Laplace (1749-1827) consolidated the subject's foundations, although it was only in the 1930s, with the works of Andrei Nikolaevich Kolmogorov (1903-1987), that the axiomatic structure of probability became definitively established.

Irénée-Jules Bienaymé (1796-1878), a French mathematician, born in Paris, continued the work of Pierre-Simon de Laplace (1749-1827) in the fields of probability and statistics. He generalized the method of square means, applied statistics to finances and demography, and enunciated the inequality that bears his name (Alencar, 2013).

Bienaymé taught in the *Académie Militaire de Saint-Cyr*, and later became professor of probability at Sorbonne and member of the *Académie des Sciences*. He also translated the works of his friend Pafnuty Lvovich Tchebychev (1821-1894), and was a critic of the Law of Large Numbers, announced by Siméon Denis Poisson (1781-1840).

Tchebychev was born in the village of Okatovo, in Russia, and was initially home-schooled by his mother and a cousin. Since he had one leg longer than the other he had a limp, which prevented him from playing with other children, so he focused mainly on studying. He graduated from Moscow University.

Considered to be the founder of Russian Mathematics, Tchebychev is known for his works in the fields of probability, statistics, and numbers theory. He took for himself the vast program initiated by Jacques Bernoulli (1654-1705), Abraham de Moivre (1667-1754), and Poisson, to announce and demonstrate in rigorous manner the limit theorems, that is, to establish the asymptotic tendencies of natural phenomena.

From this research, he established a general law of large numbers, with a new and brilliant demonstration, based on the inequality enunciated by Bienaymé, for which he also provided a new proof.

Tchebychev was the professor of two great mathematicians, Andrei Andreyevich Markov (1856-1922), known for his work in stochastic processes, the Markov chains, and Aleksandr Mikhailovich Lyapunov (1857-1918), known for developing the theory on stability of dynamic systems, as well as for his contributions in the fields of Physical Mathematics and Probability Theory.

Markov demonstrated the inequality that bears his name, in partnership with his younger brother, Vladimir Andreevich Markov (1871-1897), who was also a student of Tchebychev. His son, also named Andrei Andreevich Markov (1903-1979), was a notable mathematician, who contributed to the theory of recursive numbers and the field of Constructive Mathematics.

Markov's aptitude for Mathematics revealed itself early on and, at age 17, he presented a new method for solving ordinary linear differential equations. He studied at the University of St. Petersbourg, where he became a professor after receiving his Master's degree. After finishing his Doctorate, he was appointed for the position of extraordinary professor and elected to the Academy of Sciences.

He was removed from his functions as professor at the University of St. Petersbourg for refusing to keep tabs on students after a student strike, in 1808, and decided to retire. He was also expelled from the Russian Orthodox Church for being an atheist and for having protested against the excommunication of Leon Tolstoi. After the Revolution of 1917, he returned to teaching Probability Theory and Differential Calculus until his death, in 1922.

Nevertheless, probability calculations can be troublesome for complicated probability density functions, and even impracticable, when these functions are not available or are unknown. Hence, some mathematicians developed fundamental inequalities, that allow the determination of probabilities in an approximate manner, but with a controlled error.

8.2 Tchebychev's Inequality

If a random variable X has a finite variance, then for any $\epsilon > 0$, it is possible to establish Tchebyshev's Inequality, given as (Sveshnikov, 1968).

$$P\{|X - m_X| \geq \epsilon\} \leq \frac{\sigma_X^2}{\epsilon^2} \tag{8.1}$$

in which $\sigma_X = \sqrt{E[(X - m_X)^2]} = \sqrt{E[X^2] - E^2[X]}$ represents the standard deviation of the random variable, which is a measure of the dynamic range and helps yield the RMS value for a given signal.

By definition, the probability that $|X - m_X| \geq \epsilon$ is given by

$$P\{|X - m_X| \geq \epsilon\} = \int_{|x - m_X| \geq \epsilon} p_X(x)\,dx$$
$$= \int_{-\infty}^{m_X - \epsilon} p_X(x)\,dx + \int_{m_X + \epsilon}^{\infty} p_X(x)\,dx.$$

Since

$$\sigma_X^2 = \int_{-\infty}^{\infty} (x - m_X)^2 p_X(x)\,dx,$$

then,

$$\sigma_X^2 \geq \int_{-\infty}^{\infty} \epsilon^2 p_X(x)\,dx = \epsilon^2 \int_{-\infty}^{\infty} p_X(x)\,dx \geq \epsilon^2 \int_{|x - m_X|} p_X(x)\,dx.$$

Dividing the inequality by $\epsilon^2 > 0$, yields the desired result

$$\frac{\sigma_X^2}{\epsilon^2} \geq P\{|X - m_X| \geq \epsilon\}.$$

Example: Consider the uniform distribution, defined in the interval $[-a, a]$ by the expression

$$p_X(x) = \frac{1}{2a}[u(x + a) - u(x - a)].$$

The exact value for the probability that the random variable exceeds its mean, $m_X = 0$, by a value of ϵ is

$$P\{|x| \geq \epsilon\} = \begin{cases} 1 - \frac{\epsilon}{a}, & 0 \leq \epsilon \leq a \\ 0, & \epsilon \geq a \end{cases}$$

which results in a graph with a straight line crossing the Y-axis at 1 and the X-axis at a.

Applying Tchebychev's inequality yields

$$P\{|x| \geq \epsilon\} \leq \frac{a^2}{3\epsilon^2}$$

which results in a hyperbole, that is placed above the previous straight line. For example, when $\epsilon = a$, one has $P\{|x - m_X| \geq \epsilon\} = 0$ for the exact value, and if Tchebychev's inequality is applied, one obtains $P\{|x - m_X| \geq \epsilon\} = 1/3$, with an error of approximately 0.33. ∎

Example: For a Gaussian distribution, when the mean deviation is equal to the standard deviation, there is no advantage in using the inequality. If $\epsilon = \sigma_X$,

$$P\{|x - m_X| \geq \sigma_X\} \leq \frac{\sigma_X^2}{\sigma_X^2} = 1. \blacksquare$$

A better result is obtained for values of ϵ which are much higher than the standard deviation. For $\epsilon = 2\sigma_X$,

$$P\{|x - m_X| \geq 2\sigma_X\} \leq \frac{\sigma_X^2}{4\sigma_X^2} = 0.25.$$

Some variations of Tchebychev's inequality can be obtained (Blake, 1987), as follows:

$$P\{|x - m_X| \geq k\sigma\} \leq \frac{1}{k^2};$$

$$P\{|x - m_X| < k\sigma\} \geq 1 - \frac{1}{k^2};$$

$$P\{|x - m_X| < \epsilon\} \geq 1 - \frac{\sigma_X^2}{\epsilon^2}.$$

If $X_1, X_2, \ldots, X_n, \ldots$ is a sequence of pairwise independent random variables, whose variances are limited $V[X_i] \leq M$, for any k, then it is also possible to establish Tchebychev's Theorem,

$$\lim_{n \to \infty} P\left\{\left|\frac{1}{n}\sum_{k=1}^{n} X_k - \frac{1}{n}\sum_{k=1}^{n} m_{X_k}\right| < \epsilon\right\} = 1, \quad (8.2)$$

and Khinchin's Theorem, in case the random variables all have the same distribution, with mean value m_X,

$$\lim_{n\to\infty} P\left\{ \left| \frac{1}{n}\sum_{k=1}^{n} X_k - m_X \right| < \epsilon \right\} = 1. \tag{8.3}$$

Aleksandr Yakovlevich Khinchin (1894–1959) was a Soviet mathematician and and important contributor to the theory of probability. He is considered one of the founders of modern probability theory, and achieved important results related to limit theorems. He gave a precise definition of a stationary process and laid the foundation for the theory of such processes.

For a sequence of dependent random variables, that satisfy the condition

$$\lim_{n\to\infty} \frac{1}{n^2} V\left[\sum_{k=1}^{n} X_k \right] = 0, \tag{8.4}$$

for any small constant $\epsilon > 0$, one has

$$\lim_{n\to\infty} P\left\{ \left| \frac{1}{n}\sum_{k=1}^{n} X_k - \frac{1}{n}\sum_{k=1}^{n} m_{X_k} \right| < \epsilon \right\} = 1, \tag{8.5}$$

which is known as Markov's Theorem.

Example: It is possible to verify that if $f(x)$ is a monotonic increasing positive function, and $\mathrm{E}[f(X)] = m$ exists, then

$$P\{X > t\} \leq \frac{m}{f(t)}.$$

Considering the properties of $f(x)$, it is feasible to obtain a sequence of inequalities,

$$P\{X > t\} = \int_{x>t} f(x)\mathrm{d}x \leq \frac{1}{f(t)} \int_{x>t} f(x)p_X(x)\mathrm{d}x \tag{8.6}$$
$$\leq \frac{1}{f(t)} \int_{-\infty}^{\infty} f(x)p_X(x)\mathrm{d}x = \frac{\mathrm{E}[f(X)]}{f(t)}. \blacksquare$$

8.3 Markov's Inequality

Markov's inequality is valid only for distributions that are on the right side of the Y-axis, that is, only if $p_X(x) = 0$ when $x < 0$.

$$P\{X \geq \alpha\} \leq \frac{\mathrm{E}[X]}{\alpha}, \quad X \geq 0, \ \alpha > 0 \tag{8.7}$$

Markov's inequality can be demonstrated without great effort.

Consider the expected value defined as

$$E[X] = \int_{-\infty}^{\infty} x p_X(x) \, dx \geq \int_{\alpha}^{\infty} x p_X(x) dx = \alpha \int_{\alpha}^{\infty} p_X(x) dx = \alpha P\{x \geq \alpha\}$$

which becomes Markov's inequality if both sides are divided by α.

Example: The value of α must be greater than the variable mean, so that a non trivial value can be obtained. For the parameter $\alpha = E[X]$, $P\{x \geq m_X\} \leq \frac{E[X]}{E[X]} = 1$. ∎

8.4 Bienaymé's Inequality

Invoking once again Markov's inequality, but with a variable replacement

$$P\{Y \geq \alpha\} \leq \frac{E[Y]}{\alpha}, \; Y \geq 0, \; \alpha > 0, \tag{8.8}$$

one may notice that it continues to be valid if the following replacements are done $Y = |X - m_X|^n$ and $\alpha = \epsilon^n$.

Therefore,

$$P\{|X - m_X|^n \geq \epsilon^n\} \leq \frac{E[|X - m_X|^n]}{\epsilon^n}, \; \epsilon > 0.$$

Thereby,

$$P\{|X - m_X| \geq \epsilon\} \leq \frac{E[|X - m_X|^n]}{\epsilon^n}, \; \epsilon > 0,$$

which is Bienaymé's inequality.

8.5 Jensen's Inequality

The inequality developed by the Danish mathematician Johan Ludwig William Valdemar Jensen (1859-1925) states that, for convex functions, the expected value of $f(X)$ is greater than or equal to $f(E[X])$ (Petrov and Mordecki, 2008).

$$E[f(X)] \geq f(E[X]). \tag{8.9}$$

A continuous function, $f(x)$, is said to be convex for a given interval if a line segment between any two points on the graph of the function lies above or on top of the graph. That is equivalent to saying that the graph is above any tangent to any of its points in that interval. Another definition is that a secant line to the graph of the function is always above the graph. Therefore,

$$\lambda f(x) + (1 - \lambda)f(y) \geq f(\lambda x + (1 - \lambda)y), \text{ forall } x, y, \lambda \in R, 0 \leq \lambda \leq 1. \tag{8.10}$$

Let $f(\cdot)$ be a convex function. Then there exists a straight line that passes through the point $(\mathrm{E}[X], f(\mathrm{E}[X]))$, tangent to the graph of the function. Consider this straight line as given by $g(x) = aX + b$. At the relevant point, $f(X) = g(X)$, and for the rest, $f(X) \geq g(X)$. So,

$$f(\mathrm{E}[X]) = g(\mathrm{E}[X]) = a\mathrm{E}[X] + b,$$

but, taking in account the linear properties of the expected value,

$$a\mathrm{E}[X] + b = \mathrm{E}[aX + b] = \mathrm{E}[g(X)] \leq \mathrm{E}[f(X)],$$

as it was intended to demonstrate (Magalhães, 2006).

Jensen's inequality can be extended to cases with conditional expectation, considering two random variables, X and Y, that may or may not be dependent (Petrov and Mordecki, 2008)

$$\mathrm{E}[f(X)|Y] \geq f(\mathrm{E}[X|Y]). \tag{8.11}$$

From this result, for example, it can be shown that if X is a random variable with a finite mean, then one can show that $\mathrm{E}[X^2] \geq \mathrm{E}[X]^2$, that is, the mean square value of a random variable is always larger than or equal to the square value of the mean.

Using Jensen's inequality, it can also be proven that, for a random variable X and a given number $\alpha \in \mathbb{R}$ (Rosenthal, 2000),

$$\mathrm{E}[\max(X, \alpha)] \geq \max \mathrm{E}[X, \alpha].$$

The application of Jensen's inequality for the function $f(x) = |x|^p$, for $1 < p < \infty$, leads to

$$\mathrm{E}[|X|] \leq \left(\mathrm{E}[|X|^p]\right)^{1/p}. \tag{8.12}$$

8.6 Chernoff's Inequality

The inequality was established by Herman Chernoff (1923-), an American applied mathematician, statistician and physicist, and is better known as Chernoff's limit. It can be obtained using Markov's inequality, as follows,

$$P\{X \geq \alpha\} \leq \frac{\mathrm{E}[X]}{\alpha}, \, X \geq 0, \, \alpha > 0.$$

200 Probability Fundamental Inequalities

Replacing X by e^{tX} yields

$$P\{X \geq \alpha\} = P\{e^{tX} \geq e^{t\alpha}\} \leq \frac{\mathrm{E}[e^{tX}]}{e^{t\alpha}}. \tag{8.13}$$

This inequality is useful when there is information about the random variable, because the properties of its characteristic function, sometimes written as $P_X(t) = e^{tX}$, can be used to obtain closer limit.

A limit of this nature can be obtained for cases in which t is a real number between 0 and 1, and $X = \sum_{i=1}^{n} X_i$, in which X_i are independent random variables, such that $\mathrm{E}[X_i] = 0$ and $|X_i| \leq 1$ for any i.

Let σ be the standard deviation of X, then, using Chernoff's limit, one obtains

$$P\{X \geq \alpha\sigma\} \leq \frac{\mathrm{E}[e^{tX}]}{e^{t\alpha\sigma}}.$$

Applying the definition of the expected value,

$$\begin{aligned}
\mathrm{E}[e^{tX}] &= \sum_{k=1}^{m} p_k e^{tx_k} \\
&= \sum_{k=1}^{m} p_k \left(1 + tx_k + \frac{1}{2!}(tx_k)^2 + \frac{1}{3!}(tx_k)^3 + \cdots\right) \\
&= \sum_{k=1}^{m} p_k + t\sum_{k=1}^{m} p_k x_k + \sum_{k=1}^{m} p_k \left(\frac{1}{2!}(tx_k)^2 + \frac{1}{3!}(tx_k)^3 + \cdots\right).
\end{aligned}$$

The first sum is unitary, the second sum is the mean value of the variable X, which is considered zero, and the remaining terms can be limited by the upper bound

$$\sum_{k=1}^{m} p_k (tx_k)^2 \left(\frac{1}{2!} + \frac{1}{3!} + \cdots\right) \leq t^2 \sum_{k=1}^{m} p_k (x_k)^2.$$

But the sum to the right of the inequality is the variance of

$$\mathrm{E}[e^{tX}] \leq 1 + t^2 \mathrm{V}[X].$$

However,
$$\begin{aligned}
\mathrm{E}[e^{tX}] &= \mathrm{E}[e^{t(X_1+X_2+\cdots X_n)}] \\
&= \mathrm{E}[\prod_{i=1}^n e^{tX_i}] \\
&= \prod_{i=1}^n \mathrm{E}[e^{tX_i}], \ X_i \text{ independent} \\
&\leq \prod_{i=1}^n (1+t^2 \mathrm{V}[X]) \\
&\leq \prod_{i=1}^n e^{t^2 \mathrm{V}[X]}, \ \text{because } 1+\alpha \leq e^\alpha, \text{for } \alpha \geq 0 \\
&= e^{t^2 \sigma^2}, \ X_i \text{ independent.}
\end{aligned}$$

Substituting the previous result into Chernoff's inequality, gives

$$P\{X \geq \alpha\sigma\} \leq \frac{e^{t^2\sigma^2}}{e^{t\alpha\sigma}}$$
$$= e^{t\sigma(t\sigma-\alpha)}. \tag{8.14}$$

The optimization, to obtain the closest possible limit, leads to $t = \alpha/2\sigma$, which yields

$$P\{X \geq \alpha\sigma\} \leq e^{-\alpha^2/4}. \tag{8.15}$$

This is an interesting result, since it relates the probability of the sum of independent random variables to the Gaussian function. That is, the probability of exceeding the standard deviation by a given multiplying factor is bounded from above by the Gaussian curve, which has the factor as its parameter.

8.7 Kolmogorov's Inequality

Let $\{X_i\}$ be a sequence of independent random variables with zero mean and finite variance, given by $V[X_i] < \infty$. The probability that the absolute value of the sum, $S_n = \sum_{i=1}^n X_i$, exceeds a given value, $\alpha > 0$, obeys Kolmogorov's inequality.

$$P\{\max_{1 \leq i \leq n} |S_i| \geq \alpha\} \leq \frac{\mathrm{V}[S_n]}{\alpha^2}. \tag{8.16}$$

To demonstrate the inequality, let us form a set $A = \{\max_{1 \leq i \leq n} S_i^2 \geq \alpha^2\}$ and define the sets $A_i, 1 \leq i \leq n$ as

$$A_1 = \{S_1^2 \geq \alpha^2\};$$

$$A_2 = \{S_1^2 < \alpha^2, S_2^2 \geq \alpha^2\};$$

$$A_i = \{S_1^2 < \alpha^2, S_2^2 < \alpha^2, \ldots, S_{i-1}^2 < \alpha^2, S_i^2 \geq \alpha^2\}, \text{ for } 2 \leq i \leq n.$$

These sets, which are mutually disjoint, represent the first occurrence in which S_i^2 exceeds α^2. Thus,

$$A = \bigcup_{i=1}^{n} A_i \text{ and } I_A(s) = \sum_{i=1}^{n} I_{A_i}(s).$$

in which $I_A(s)$ is the indicator function, defined as

$$I_A(s) = 1, \text{ if } s \in A, \text{ and}$$
$$I_A(s) = 0, \text{ if } s \notin A. \qquad (8.17)$$

Since the mean of the sum is assumed to be zero, the variance is equal to the mean square, that is,

$$V[S_n^2] = E[S_n^2] = \int_{-\infty}^{\infty} s_n^2 dP_{S_n}(s_n).$$

However, since $S_* = \max_{1 \leq i \leq n} |S_i| \geq \alpha$ is the established condition, then $S_*^2 = \max_{1 \leq i \leq n} S_i^2 \geq \alpha^2$, that is,

$$E[S_n^2] \geq \alpha^2 \int_0^{\infty} dP_{S_*^2}(s_*^2) \geq \alpha^2 \int_{\alpha^2}^{\infty} dP_{S^2}(s^2)$$
$$= \alpha^2 \int I_A(s) dP_S(s) = \alpha^2 P\{\max_{1 \leq i \leq n} |S_i| \geq \alpha\},$$

which verifies the inequality, in which $P_S(s)$ is the cumulative distribution function of the random variable S.

8.8 Schwarz' Inequality

Schwarz's inequality, also known as the Cauchy-Schwarz inequality, is widely used to demonstrate theorems and other important results. Initially, the inequality for sums was published by Augustin-Louis Cauchy (1789-1857), a French mathematician, in 1821. The inequality for integrals was firstly established by Viktor Yakovlevich Bunyakovsky (1804-1889), in St. Petersbourg, Russia, in 1859, and rediscovered by Hermann Amandus Schwarz (1843-1921), in Germany, in 1888.

$$\mathrm{E}\left[(XY)\right] \leq \sqrt{\mathrm{E}\left[(X^2)\right]}\sqrt{\mathrm{E}\left[(Y^2)\right]}. \tag{8.18}$$

To establish Schwarz's inequality, it is sufficient to take the mathematical tautology

$$\mathrm{E}\left[(X + \alpha Y)^2\right] \geq 0.$$

Expanding the binomial and minimizing for α, which is done by making the derivative equal to zero,

$$\frac{\partial}{\partial \alpha}\left(\mathrm{E}\left[X^2\right] + 2\alpha \mathrm{E}\left[XY\right] + \alpha^2 \mathrm{E}\left[Y^2\right]\right) = 0,$$

results in

$$\alpha = -\frac{\mathrm{E}\left[XY\right]}{\mathrm{E}\left[Y^2\right]}.$$

Replacing α in the tautology yields

$$\left(\mathrm{E}\left[X^2\right] - 2\frac{\mathrm{E}\left[XY\right]}{\mathrm{E}\left[Y^2\right]}\mathrm{E}\left[XY\right] + \left(\frac{\mathrm{E}\left[XY\right]}{\mathrm{E}\left[Y^2\right]}\right)^2 \mathrm{E}\left[Y^2\right]\right) \geq 0,$$

which can be simplified to

$$\mathrm{E}\left[X^2\right] - \frac{\mathrm{E}^2\left[XY\right]}{\mathrm{E}\left[Y^2\right]} \geq 0.$$

Therefore,

$$\mathrm{E}\left[X^2\right] \geq \frac{\mathrm{E}^2\left[XY\right]}{\mathrm{E}\left[Y^2\right]},$$

and finally

$$\mathrm{E}\left[(XY)\right] \leq \sqrt{\mathrm{E}\left[(X^2)\right]}\sqrt{\mathrm{E}\left[(Y^2)\right]}.$$

8.9 Hölder's Inequality

The inequality studied by Otto Ludwig Hölder (1859-1937), a German mathematician, who studied with Leopold Kronecker (1823-1891), Karl Theodor Wilhelm Weierstrass (1815-1897) and Ernst Eduard Kummer (1810-1893), is used to establish the result for high-order spaces.

Hölder's inequality, generally used to establish the integrability of the product XY, is given by

$$E[|XY|] \leq \left(E[|X|^p]\right)^{1/p} \cdot \left(E[|Y|^q]\right)^{1/q}. \tag{8.19}$$

When $p = q = 2$, Hölder's inequality becomes Schwarz's inequality. When X and Y are integrable random variables, then $X + Y$ is also integrable, which is a consequence of the triangle inequality

$$|X + Y| \leq |X| + |Y|.$$

Besides,

$$E[|X + Y|] \leq E[|X|] + E[|Y|]. \tag{8.20}$$

8.10 Lyapunov's Inequality

Aleksandr Mikhailovich Lyapunov (1857-1918) studied the inequality that holds his name, and made various contributions to the fields of Mathematics, Physics, and Probability Theory.

Assume that $0 < \alpha \leq \beta$, making $p = \beta/\alpha$, $q = \beta/(\beta - \alpha)$, $Y = 1$ and replacing X by $|X|^\alpha$ in Hölder's inequality 8.19, one obtains Lyapunov's inequality (Billingsley, 1995),

$$\left(E[|X|^\alpha]\right)^{1/\alpha} \leq \left(E[|X|^\beta]\right)^{1/\beta}, \; 0 < \alpha \leq \beta. \tag{8.21}$$

8.11 Minkowsky's Inequality

Let $p \in [1, \infty)$ and X and Y such that $E[|X|^p] < \infty$ and $E[|Y|^p] < \infty$, so the inequality assigned to Hermann Minkowski (1864-1909), a Lithuanian mathematician, is given by

$$E^{1/p}[|X + Y|^p] \leq \left(E[|X|^p]\right)^{1/p} + \left(E[|Y|^p]\right)^{1/p}. \tag{8.22}$$

To demonstrate the inequality, it is sufficient to assume $p > 1$, taking into account that the case in which $p = 1$ was already treated in the section about Jensen's inequality. It can be seen that the function

$$f(x) = (1 + x)^p - 2^{p-1}(1 + x)^p$$

reaches its maximum when $x = 1$, and $f(1) = 0$, that is,

$$(1 + x)^p \leq 2^{p-1}(1 + x)^p,$$

which becomes, after the replacement, $x = |Y|/|X|$,

$$\begin{aligned}|X+Y|^p &\leq (|X|+|Y|)^p = |X|^p(1+x)^p \\ &\leq |X|^p 2^{p-1}(1+x)^p, \\ &= 2^{p-1}(|X|^p+|Y|^p)\end{aligned}$$

The integrability of $|X+Y|^p$ is proven. Let

$$q = \frac{p}{p-1}, \text{ such that, } \frac{1}{p}+\frac{1}{q}=1.$$

Applying the inequality for $Y = |U+V|^{p-1}$ and $X = U$, then $V = U$, and one obtains

$$\mathrm{E}\left[|U|\cdot|U+V|^{p-1}\right] \leq \mathrm{E}^{1/p}\left[|U|^p\right]\cdot \mathrm{E}^{1/p}\left[|U+V|^p\right]$$

and

$$\mathrm{E}\left[|V|\cdot|U+V|^{p-1}\right] \leq \mathrm{E}^{1/p}\left[|V|^p\right]\cdot \mathrm{E}^{1/p}\left[|U+V|^p\right].$$

Therefore, considering that

$$|U+V|^p = |U+V|\cdot|U+V|^{p-1} \leq |U|\cdot|U+V|^{p-1}+|V|\cdot|U+V|^{p-1},$$

one has

$$\mathrm{E}\left[|U+V|^p\right] \leq \mathrm{E}^{1/p}\left[|U+V|^p\right]\cdot\left(\mathrm{E}^{1/p}\left[|U|^p\right]+\mathrm{E}^{1/p}\left[|V|^p\right]\right). \quad (8.23)$$

Dividing by $\mathrm{E}^{1/q}\left[|U+V|^p\right]$ results in Minkowsky's inequality.

8.12 Fatou's Lemma

Pierre Joseph Louis Fatou (1878–1929), a French mathematician and astronomer, known for major contributions to several branches of analysis, derived the lemma that bears his name for his thesis *Séries Trigonométriques et Séries de Taylor*, in 1906, which was the first application of the Lebesgue integral to concrete problems of analysis (Fatou, 1906).

For a sequence of functions $\{f_n\}$, which converge pointwise to some limit function f, it is not always true that

$$\int \lim_{n\to\infty} f_n = \lim_{n\to\infty}\int f_n.$$

Fatou's Lemma is a result in the theory of Lebesgue integration which answers the question of when the limit and the integration are allowed to commute. That is, Fatou¿s Lemma is an inequality that demonstrates a relationship between the Lebesgue integral of a sequence of functions' limit inferior and the limit inferior of integrals of these functions. In other words, Fatou's Lemma indicates the best that can be done if there are no restrictions on the functions.

Given a measure space $(\Omega, \mathcal{F}, \mu)$ and a set $X \in \mathcal{F}$, let $\{f_n\}$ be a sequence of $(\mathcal{F}, \mathcal{B}_{\mathbb{R}+})$-measurable non-negative functions $f_n : X \to [0, +\infty]$. Define the function $f : X \to [0, +\infty]$ by setting

$$f(x) = \liminf_{n \to \infty} f_n(x),$$

for every $x \in X$.

Then f is $(\mathcal{F}, \mathcal{B}_{\mathbb{R}+})$-measurable, and also

$$\int_X f \, d\mu \leq \liminf_{n \to \infty} \int_X f_n \, d\mu, \qquad (8.24)$$

in which the integrals may be infinite. Fatou's Lemma remains true if the assumptions hold μ-almost everywhere, and does not require the monotone convergence theorem, but it can be used to provide a short proof.

First, consider that the properties of $g_n(x) = \inf_{k \geq n} f_k(x)$ satisfy the following assertions:

1. The sequence $\{g_n(x)\}_n$ is pointwise non-decreasing at any x,

2. The functions obey $g_n \leq f_n$, $\forall n \in \mathbb{N}$.

Because

$$f(x) = \liminf_{n \to \infty} f_n(x) = \lim_{n \to \infty} \inf_{k \geq n} f_k(x) = \lim_{n \to \infty} g_n(x),$$

one concludes that f is measurable, and it is possible to write

$$\int_X f \, d\mu = \int_X \lim_n g_n \, d\mu$$

Using the Monotone Convergence Theorem and Property 1, the limit operation and the integral may be interchanged, yielding

$$\int_X f\, d\mu = \lim_n \int_X g_n\, d\mu$$
$$= \liminf_n \int_X g_n\, d\mu$$
$$\leq \liminf_n \int_X f_n\, d\mu,$$

Property 2 was used in the last step.

Example: For the following sequence of functions the inequality becomes strict,

$$f_n(x) = \begin{cases} 0 & \text{if } x \in [-n, n], \\ 1 & \text{otherwise.} \end{cases}$$

8.13 About Arguments and Proofs

Timon of Phlius (320-230 a.C.), shown in Figure 8.1, a skeptical philosopher of Ancient Greece, showed that every argument, or proof, proceeded from premises that the argument itself does not establish. If one tries to establish the truth of these premises using other arguments or proofs, then it will be based on assumed premises, and so on, *ad infinitum*. Therefore, there is no way to establish a definitive base of certainty.

Timon was a disciple of Pyrrho (c. 365-270 BC), who created a philosophical school in which the members were known as skeptics. The skeptics sought to put all truths to the test, systematically and thoroughly. Pyrrho served as a soldier in the army of Alexander the Great (356-323 a.C.), was an apprentice of Aristotle (384-322 a.C.), one of the most influential philosophers of all times, and built a school of philosophy known as the Lyceum.

In the words of Timon, what a valid argument proves is that its conclusions stem from its premises, but this is not the same as proving that these conclusions are true. All valid argument is of the form "If the premise P is true, then the question Q must be true." The argument itself does not prove this truth, because it assumed the truth *a priori*, and to assume what is to be proven generates a vicious cycle. Thus, all proof rests on unproven assumptions, which is true for any science, including Logic and Mathematics, or life in general.

Figure 8.1 Timon of Phlius, Ancient Greek skeptic philosopher. Source: Thomas Stanley, 1655, The History of Philosophy, Public Domain, https://commons.wikimedia.org/w/index.php?curid=8722615

8.14 Problems

1. Use Tchebyshev's Inequality to obtain an estimate for the probability that a random variable with exponential distribution, $p_X(x) = \alpha e^{-\alpha x} u(x)$, exceeds its mean by a value of ϵ, and compare with the value calculated using the distribution.

$$P\{|X - m_X| \geq \epsilon\} \leq \frac{\sigma_X^2}{\epsilon^2},$$

in which $\sigma_X = \sqrt{E[(X - m_X)^2]}$, represents the standard deviation of the random variable.

2. Markov's inequality is valid for the distribution given in the previous question. Determine an estimate.

$$P\{X \geq \alpha\} \leq \frac{E[X]}{\alpha}, \; X \geq 0, \; \alpha > 0.$$

3. Use Jensen's inequality to prove that if X is a random variable with a finite mean, then $E[X^2] \geq E[X]^2$, that is, the mean square value of a

random variable is always larger than or equal to the square value of the mean.

4. Explain Hölder's inequality for the cases $p = q = 1$ and $p = q = 2$.
$$\mathrm{E}[|XY|] \leq \left(\mathrm{E}[|X|^p]\right)^{1/p} \cdot \left(\mathrm{E}[|Y|^q]\right)^{1/q}.$$

5. Use Hölder's inequality to derive Lyapunov's inequality,
$$\left(\mathrm{E}[|X|^\alpha]\right)^{1/\alpha} \leq \left(\mathrm{E}[|X|^\beta]\right)^{1/\beta}, \ 0 < \alpha \leq \beta.$$

6. Explain Minkowsky's inequality for the case $p = 1$.
$$\mathrm{E}^{1/p}[|X+Y|^p] \leq \left(\mathrm{E}[|X|^p]\right)^{1/p} + \left(\mathrm{E}[|Y|^p]\right)^{1/p}.$$

9

Convergence and the Law of Large Numbers

"Mathematics is the science of what is clear by itself."
Carl Jacobi

9.1 Forms of Convergence in Probability Theory

The Mathematics of the nineteenth century was very creative. However, in some areas, such as Integral Calculus, it lacked the necessary rigor. Number theory is another example, especially regarding the convergence limits of sequences. Karl Weierstrass (1815-1897), a high school teacher in a small German town, was responsible for introducing exactitude in the analytical arguments of the time and, despite his proverbial lack of interest in publishing his results, he became known as the father of modern analysis (Dunham, 2005).

Weierstrass was the first to recognize the dichotomy between point-wise convergence and uniform convergence, and to show that the limit of the integral of a sequence of functions was not always equal to the integral of the limit of that sequence, at least for the Riemann integral.

9.2 Types of Convergence

In the field of probability, there are many definitions of convergence of random variables, according to the specific application. The convergence of a sequence of random variables for a given limit is conceptually important in the Theory of Probability, with applications in Statistics and Stochastic Processes.

9.2.1 Convergence in Probability

Consider a probability space $(\Omega, \mathcal{F}, \mathcal{P})$, in which Ω represents the sample space and $\mathcal{F}$ is the algebra in which the probability measure P is applied, and let $X, X_1, X_2, \ldots$ be random variables defined in this space.

It is said that a sequence of random variables, $\{X_n\}$, converges in probability to the random variable X, that is, $X_n \xrightarrow{P} X$, if for every $\epsilon > 0$,

$$P\{|X_n - X| \geq \epsilon\} \to 0, \tag{9.1}$$

as $n \to \infty$.

The equivalent definition is as follows. For every $\epsilon > 0$,

$$P\{|X_n - X| < \epsilon\} \to 1, \tag{9.2}$$

when $n \to \infty$. Furthermore, convergence in probability implies convergence in distribution.

9.2.2 Almost Sure Convergence

A sequence of random variables, X_n, converges almost surely or almost everywhere or with probability 1 or converges strongly towards the random variable X, which can be written as $X_n \xrightarrow{a.s.} X$, if $X_n(\omega) \to X(\omega)$ for all the points $\omega \in \Omega$.

In mathematical notation

$$P\{\lim_{n \to \infty} X_n = X\} = 1, \tag{9.3}$$

or

$$P\{\omega \in \Omega : \lim_{n \to \infty} X_n(\omega) = X(\omega)\} = 1. \tag{9.4}$$

In real analysis, the convergence in probability corresponds to the convergence in measure. The almost sure convergence corresponds to the convergence at almost every point. Since the convergence at almost every point implies convergence in measure, then the almost sure convergence implies convergence in probability.

Assume that $X_n \xrightarrow{a.s.} X$. Then, for a given $\epsilon > 0$, for each $\omega \in \Omega$, with the exception of a result with zero probability, there exists $n = f(\epsilon, \omega)$ such that for every $k \geq n$, there is an $|X_k(\omega) - X(\omega)| < \epsilon$.

That is, given $\epsilon > 0$, then

$$P\{\bigcup_{n=1}^{\infty} \bigcap_{n=1}^{\infty} (|X_n(\omega) - X(\omega)| < \epsilon)\} = 1. \tag{9.5}$$

The complement of the set is given by

$$P\{\bigcap_{n=1}^{\infty} \bigcup_{n=1}^{\infty} (|X_n(\omega) - X(\omega)| \geq \epsilon)\} = 0. \tag{9.6}$$

The sequence of sets

$$D_n = \bigcup_{k=n}^{\infty} (|X_n(\omega) - X(\omega)| \geq \epsilon)$$

meets the condition $D_1 \supset D_2 \supset D_3 \cdots$, that is, it is a decreasing sequence. Assuming that Expression 9.6 is verified, then

$$\lim_{n \to \infty} P\{D_n\} = P\{\bigcap_{n=1}^{\infty} D_n\} = 0.$$

Therefore,

$$P\{|X_n - X| \geq \epsilon\} \leq P\{\bigcup_{k=n}^{\infty} |X_k - X| \geq \epsilon)\} = P\{D_n\} \to 0,$$

if $n \to \infty$, for any $\epsilon > 0$. As a consequence, $X_n \xrightarrow{P} X$, which concludes the demonstration.

9.2.3 Sure Convergence

A sequence of random variables $\{X_n\}$ defined in the same probability space converges surely or everywhere or point-wise towards X when

$$\lim_{n \to \infty} X_n(\omega) = X(\omega), \forall \omega \in \Omega, \tag{9.7}$$

in which Ω is the sample space in which the random variables are defined.

This is the concept of point-wise convergence of a sequence of functions extended to a sequence of random variables, that represent the functions.

Sure convergence implies all the other kinds of convergence, but there is no advantage in using sure convergence compared to using almost sure convergence. The difference between the two only exists in sets with probability zero. This is why the concept of sure convergence of random variables is rarely used.

9.2.4 Convergence in Distribution

Consider a sequence of cumulative probability functions, that can be written as $P_{X_1}(x), P_{X_2}(x), P_{X_3}(x), \ldots$ relative to the random variables $X_1, X_2, X_3 \ldots$, and let $P_X(x)$ be the cumulative function relative to the random variable X. It is said that a sequence $\{X_n\}$ converges in distribution to the random variable

$$\lim_{n \to \infty} P_{X_n}(x) = P_X(x), \qquad (9.8)$$

for every number x for which $P_X(x)$ is continuous.

Convergence in distribution can be denoted by a letter placed over the arrow $X_n \xrightarrow{d} X$. It is also known as weak convergence, since it is implied by all the other types of convergence and does not imply any of them. However, it is a useful definition for the convergence of random variables and is used in the Central Limit Theorem.

Convergence in probability implies convergence in distribution, that is, if $X_n \xrightarrow{P} X$, then $X_n \xrightarrow{d} X$.

9.2.5 Convergence in Mean of Order r

Assume that the random variables $X_1, X_2, X_3, \ldots$ have finite moment of order $r \geq 1$. It is said that the sequence $\{X_n\}$ converges in the r-th mean or in the L^r-norm for X, if

$$\lim_{n \to \infty} \mathrm{E}\left(|X_n - X|^r\right) = 0, \qquad (9.9)$$

when $n \to \infty$, in which the operator E denotes the expected value. This type of convergence is often denoted by adding the letter r over an arrow indicating convergence.

Convergence of the r-th mean establishes that the expected value of the r-th power of the difference between X_n and X converges to zero, as the sequence grows. There is no pattern to represent this convergence, but it can be represented with the addition of the symbol r over the arrow $X_n \xrightarrow{r} X$. For $r = 1$, it is said that X_n converges in mean to X.

9.2.6 Convergence in Mean

Assume that the random variables $X_1, X_2, X_3, \ldots$ have finite moment or first order. It is said that the sequence $\{X_n\}$ converges in in mean or has a mean convergence, if (Gubner, 2006)

$$\lim_{n \to \infty} \mathrm{E}\left(|X_n - X|\right) = 0, \quad (9.10)$$

when $n \to \infty$, in which the operator E denotes the expected value.

Convergence in mean establishes that the expected value of the modulus of the difference between X_n and X converges to zero, as the sequence grows.

9.2.7 Convergence in Mean Square

For $r = 2$, in the Formula 9.9, it is said that X_n converges in mean square to X, by writing

$$\lim_{n \to \infty} \mathrm{E}\left(|X_n - X|^2\right) = 0, \quad (9.11)$$

when $n \to \infty$, in which the operator E denotes the expected value. Sometimes, it is called convergence in mean, and is represented by $\underset{n \to \infty}{\mathrm{l.i.m.}} X_n = X$.

Convergence in the r-th moment, with $r > 0$, implies convergence in probability, from Markov's inequality. If $r > s \geq 1$, convergence in the r-th mean implies convergence in the s-th power, therefore, convergence in mean square implies convergence in mean.

Example: Consider that the random variables $X_1, X_2, X_3, \ldots$ are uncorrelated with mean μ and variance σ^2. Then, the sample mean, defined in the following, converges in mean square to μ.

$$\bar{X}_N = \frac{1}{N} \sum_{n=1}^{N} X_n.$$

It is possible to write

$$\mathrm{E}[|\bar{X}_N - \mu|^2] = \frac{1}{N^2} \mathrm{E}\left[\left(\sum_{n=1}^{N}(X_n - \mu)\right)\left(\sum_{k=1}^{N}(X_k - \mu)\right)\right].$$

Rearranging the terms in the equation, one obtains

$$\mathrm{E}[|\bar{X}_N - \mu|^2] = \frac{1}{N^2} \sum_{n=1}^{N} \sum_{k=1}^{N} \mathrm{E}\left[(X_n - \mu)(X_k - \mu)\right].$$

216 *Convergence and the Law of Large Numbers*

Recalling that X_n and X_k are uncorrelated, the expectations are null when $n \neq k$, which gives

$$E[|\bar{X}_N - \mu|^2] = \frac{1}{N^2} \sum_{n=1}^{N} \sum_{k=1}^{N} E\left[(X_n - \mu)^2\right] = \frac{N\sigma^2}{N^2} = \frac{\sigma^2}{N},$$

which converges to zero when $N \to \infty$.

9.2.8 Convergence in Measure

There is relation between a sequence of measurable functions $\{f_n\}$ and a measurable function f. Consider the measure space $(\Omega, \mathcal{F}, \mu)$ and let f and $f_1, f_2, \ldots$ be measurable functions of Ω in $\mathbb{R}$, with finite values almost everywhere. For each $\epsilon > 0$ and each n, consider the set (Taylor, 1985)

$$F_n(\epsilon) = \{x : |f_n(x) - f(x)| \geq \epsilon\}. \tag{9.12}$$

This set is measurable. If $\mu(F_n(\epsilon)) \to 0$, when $n \to \infty$, for each $\epsilon > 0$, it is said that $\{f_n\}$ converges in measure to f. The notation $f_n \xrightarrow{\mu} f$ is used to express this convergence.

9.3 Relationships Between the Types of Convergence

Some modes of convergence imply others. For example, it can be seen that if $X_n \xrightarrow{r} X$, then $X_n \xrightarrow{P} X$, which can be proven by applying Tchebychev's inequality

$$P\{|X_n - X| \geq \epsilon\} = P\{|X_n - X|^r \geq \epsilon^r\} \leq \frac{1}{\epsilon^r} E\left(|X_n - X|^r\right),$$

for any $\epsilon > 0$.

If $E\left(|X_n - X|^r\right) \to 0$, then the right-hand side of the inequality tends to zero and, therefore, $P\{|X_n - X| \geq \epsilon\} \to 0$, for every $\epsilon > 0$, that is, $X_n \xrightarrow{P} X$.

Example: Given the sequences of random variables X_n and Y_n, if X_n converges in mean square to X, and Y_n converges in mean square to Y, show that the sum of the random variables, $X_n + Y_n$, converges in mean to $X + Y$.

Consider that

$$E\left[((X_n + Y_n) - (X + Y))^2\right] = E\left[((X_n - X) + (Y_n - Y))^2\right]$$

$$= \mathrm{E}\left[(X_n - X)^2\right] + \mathrm{E}\left[(Y_n - Y)^2\right]$$
$$+ 2\mathrm{E}\left[(X_n - X)(Y_n - Y)\right].$$

The first two terms tend to zero, because $X_n \to X$ and $Y_n \to Y$ in mean square. All that is left to show is that the last term also tends to zero. For this objective, Schwarz's inequality can be applied

$$\mathrm{E}\left[(XY)\right] \leq \sqrt{\mathrm{E}\left[(X^2)\right]}\sqrt{\mathrm{E}\left[(Y^2)\right]}. \tag{9.13}$$

Applying the inequality to the third term yields

$$\mathrm{E}\left[((X_n + Y_n) - (X + Y))^2\right] \leq \mathrm{E}\left[(X_n - X)^2\right] + \mathrm{E}\left[(Y_n - Y)^2\right]$$
$$+ 2\sqrt{\mathrm{E}\left[(X_n - X)^2\right]}\sqrt{\mathrm{E}\left[(Y_n - Y)^2\right]}.$$

Therefore,

$$\mathrm{E}\left[((X_n + Y_n) - (X + Y))^2\right] = \left(\sqrt{\mathrm{E}\left[(X_n - X^2)\right]} + \sqrt{\mathrm{E}\left[(Y_n - Y^2)\right]}\right)^2.$$

The last term of the equation tends to zero, when $n \to \infty$, because $X_n \to X$ and $Y_n \to Y$ in mean square. ∎

9.4 Weak Law of Large Numbers

Bernoulli produced the first demonstration for the Law of Large Numbers, on which he worked for 20 years until he obtained a sufficiently rigorous mathematical proof, that was published in his book, *Ars Conjectandi*, and became known, at the time, as the "Bernoulli Theorem". In 1835, Poisson, called it the "Law of Large Numbers," the name by which it came to be referred as afterward.

Many mathematicians contributed to the refinement of the deduction, including Tchebyshev, Markov, Borel, and Kolmogorov. The research led to two forms of the Law of Large Numbers, namely the weak law and the strong law, that define different modes to represent the convergence of probability observed for real life probability. Bernoulli's treaty about the theory of probabilities also presented applications to actuarial calculus and statistics, and was only published in 1713, after his death.

The Law of Large Numbers determines conditions for convergence of sequences of random variables. The Weak Law of Large Numbers is expressed as follows. Assume that $X_1, X_2, \ldots, X_N$ is a sequence of independent random variables, with identical probability distribution functions, with means $E(X_i) = \mu$, and with variances $V(X_i) = \sigma^2 < \infty$.

The sample mean of a variable is defined as (Blake, 1987)

$$S_N = \frac{1}{N}\sum_1^N X_i. \tag{9.14}$$

Because the variables are independent,

$$E[X_i] = \mu, \ V[X_i] = \sigma^2/N.$$

Therefore, by applying Tchebychev's inequality to the random variable S_N, it can be shown that

$$P\{|S_N - \mu| \geq \epsilon\} \leq \frac{\sigma_X^2}{N\epsilon^2}, \epsilon > 0. \tag{9.15}$$

Therefore, the mean sample approaches the statistical mean, as much as desired, as the number of samples considered increases. In fact, it is not necessary that the random variables have the same distribution.

Example: how many samples of a random variable should be taken so as to obtain a probability of at least 0.9 that the sample mean does not divert from the exact value by over $\sigma_X/4$?

Using the Inequality 9.15, it can be seen that,

$$P\{|S_N - \mu| \geq \frac{\sigma_X}{4}\} = 1 - P\{|S_N - \mu| < \frac{\sigma_X}{4}\} \leq \frac{\sigma_X^2}{N(\epsilon^2/16)},$$

that results in

$$P\{|S_N - \mu| < \frac{\sigma_X}{4}\} \geq 1 - \frac{\sigma_X^2}{N(\sigma^2/16)},$$

that is,

$$P\{|S_N - \mu| < \frac{\sigma_X}{4}\} \geq 1 - \frac{16}{N}.$$

For the probability of the event to be at least 0.9, the number of samples to be taken N must be at least $N \geq 16/0.1 = 160$. ∎

If the probability distribution was known, this number could be much smaller. However, the previous result is valid for any distribution, which can be useful in many practical cases.

Example: assume that the number of packets that arrive at a server in t seconds is a random variable $N(t)$ with a Poisson distribution and mean λt.

What is the limit for the probability that $|N(t)/t - \lambda|$ exceeds ϵ, that is, that the packet rate throughput time becomes superior to the mean rate.

For this,

$$P\left(\left|\frac{N(t)}{t} - \lambda\right| \geq \epsilon\right) = P(|N(t) - \lambda t| \geq \epsilon t)$$

$$\leq \frac{V[N(t)]}{(\epsilon t)^2}, \text{ using Tchebychev's Inequality}$$

$$= \frac{\lambda t}{(\epsilon t)^2} = \frac{\lambda}{\epsilon^2 t}. \blacksquare$$

9.5 Strong Law of Large Numbers

The strong Law of Large Numbers states that, assuming $X_1, X_2, \ldots$ a sequence of independent random variables, with the same mean μ, finite variance σ^2 and fourth centered moment also finite $E[(X_i - \mu)^4] \leq \alpha$, then

$$P\{\lim_{n \to \infty} \frac{1}{n} \sum_{1}^{n} X_n = \mu\} = 1, \tag{9.16}$$

that is, the partial means converge almost surely, or with probability 1, to the set's mean, μ.

The law's demonstration uses the following result about almost sure convergence. Let $X_1, X_2, \ldots$ be random variables and assume that, for each $\epsilon > 0$, $P\{|X_n - X| \geq \epsilon\} = 0$, infinitely often (i.o.). Therefore $P\{X_n \to X\} = 1$, that is, the sequence $\{X_n\}$ converges to X almost surely (Rosenthal, 2000).

It is assumed, as a simplification, that the mean is zero, that is, $\mu = 0$. In case it is not, it suffices to replace X_i with $X_i - \mu$ in the following deduction.

Consider the sum $S_n = X_1 + X_2 + \cdots + X_n$, and the calculation of its fourth moment $E[S_n^4]$. The expansion of S_n^4 contains many terms of the form $X_i X_j X_k X_l$, for distinct i, j, k, l, and all of the terms have expected value equal to zero. It also contains terms of the type $X_i X_j (X_k)^2$ and $X_i (X_j)^3$, all with expected values equal to zero.

The terms that do not disappear are those n of the form $(X_i)^4$ and those $3n(n-1)$ of the form $(X_i)^2 (X_j)^2$, with $i \neq j$ and, as required, $E[(X_i)^4] \leq \alpha$. Besides this, if $i \neq j$, then $(X_i)^2$ and $(X_j)^2$ are independent , which leads to $E[(X_i)^2 (X_j)^2] = E[(X_i)^2] \cdot E[(X_j)^2] = \sigma^4$.

Therefore, it can be seen that $E[S_n^4] \leq n\alpha + 3n(n-1)\sigma^4 \leq \beta n^2$, in which $\beta = \alpha + 3\sigma^4$. However, according to Markov's inequality, for any $\epsilon > 0$,

$$P\{|S_n| \geq n\epsilon\} = P\{|S_n|^4 \geq n^4\epsilon^4\} \leq \frac{E[S_n^4]}{n^4\epsilon^4} \leq \frac{\beta n^2}{n^4\epsilon^4} = \frac{\beta\epsilon^{-4}}{n^2}.$$

As long as $\sum_{n=1}^{\infty} 1/n^2 < \infty$, then $P\{|S_n| \geq n\epsilon\} = 0$, infinitely often, that is, $P\{|S_n|/n \geq \epsilon\} = 0$, infinitely often, and S_n/n converges to the zero mean almost surely. This last result uses Borel-Cantelli's Lemma (Rosenthal, 2000).

For example, in the case of a digital signal, with a sequence of independent random pulses, with values from the set $A = \{-1, 1\}$, the result indicates that the signal mean converges to zero as the number of pulses tends to infinity, $n \to \infty$.

Example: let $X_1, X_2 \ldots$ be a sequence of independent integer valued random variables, and let N be an integer valued random variable independent of X_i, for any i. Find the mean of the sum of random number of variables, given by

$$S = \sum_{i=1}^{N} X_i.$$

It should be noticed that the conditional expected value is given by

$$E[S|N=n] = E\left[\sum_{i=1}^{n} X_i\right] = nE[X],$$

thus,

$$E[S] = E[E[S|N]] = E[NE[X]] = E[N] \cdot E[X],$$

since $E[X]$ is constant.

To calculate the variance, consider that $E[S^2] = E[E[S^2|N]]$, which requires the determination of

$$E[S^2|N=n] = E\left[\sum_{i=1}^{n} X_i \sum_{k=1}^{n} X_k\right] = \sum_{i=1}^{n}\sum_{k=1}^{n} E[X_i X_k].$$

Calculating, considering that $E[X_i X_k] = E[X^2]$, if $i = k$, and that $E[X_i X_k] = E^2[X]$, if $i \neq k$, because the variables are independent, yields

9.5 Strong Law of Large Numbers

$$E[S^2|N=n] = nE[X^2] + n(n-1)E^2[X].$$

Therefore,

$$\begin{aligned}E[S^2] &= E\left[NE[X^2] + N(N-1)E^2[X]\right] \\ &= E[N]\cdot E[X^2] + E[N^2]\cdot E[X^2] - E[N]\cdot E^2[X].\end{aligned}$$

From the definition of variance,

$$\begin{aligned}V[S] &= E[S^2] - E^2[S] \\ &= E[N]\cdot E[X^2] + E[N^2]\cdot E[X^2] - E[N]\cdot E^2[X] - E^2[N]\cdot E^2[X] \\ &= E[N]\cdot V[X] + V[N]\cdot E^2[X]. \blacksquare\end{aligned} \quad (9.17)$$

Example: consider once again the sequence of independent integer valued random variables $X_1, X_2 \ldots$, and N an integer valued random variable independent of X_i, for any i, and the sum of a number of random variables given by

$$S = \sum_{i=1}^{N} X_i.$$

Show that the characteristic function of S is given by

$$P_S(\omega) = E[e^{-j\omega S}] = P_N\left(j\ln\left(P_X(\omega)\right)\right).$$

It can be verified that

$$E[e^{-j\omega S}|N=n] = E\left[e^{-j\omega \sum_{i=1}^{n} X_i}\right] = \prod_{i=1}^{n} E\left[e^{-j\omega X_i}\right] = P_X^n(\omega),$$

because the variables are identically distributed.

$$\begin{aligned}P_S(\omega) = E[e^{-j\omega S}] &= E\left[E[e^{-j\omega S}|N]\right] \\ &= E\left[P_X^N(\omega)\right] \\ &= E\left[e^{(\ln P_X(\omega))N}\right] \\ &= E\left[e^{\ln P_X(\omega)N}\right].\end{aligned}$$

Making the substitution $\ln P_X(\omega) = -j\varphi$, yields

$$P_S(\omega) = E\left[e^{-j\varphi N}\right].$$

From the definition of the characteristic function, this can be finally reduced to

$$P_S(\omega) = P_N\left(j \ln\left(P_X(\omega)\right)\right), \qquad (9.18)$$

because $\varphi = j \ln P_X(\omega)$. ∎

9.6 Central Limit Theorem

The Central Limit Theorem is a fundamental result of Statistics, for it states that when the sample size becomes larger, the sample distribution of the mean approaches that of a normal distribution.

The initial version of the theorem was proposed by the French mathematician Abraham de Moivre (1667-1754), who used the Gaussian function to approximate the distribution of the number of results from several coin tosses, in an article published in 1733.

The postulate was forgotten until another French mathematician, Pierre Simon de Laplace (1749-1827), inserted it in his book *Théorie Analytique des Probabilités*, published in 1812. Laplace generalized the result and found an approximation of the binomial distribution from the normal distribution.

Laplace's discovery received little attention, and only at the end of the 19th century the importance of the Central Limit Theorem, was realized. In 1901, the Russian mathematician Aleksandr Mikhailovich Lyapunov (1857-1918) determined more general conditions for the theorem. The expression "Central Limit Theorem" was coined by George Pólya (1887-1985), in 1920, in the title of an article. He referred to the theorem as central due to its importance in the Theory of Probability.

9.6.1 Demonstration of the Theorem

To prove the theorem, consider that $X_1, X_2, \ldots, X_N$ is a sequence of independent random variables, with identical probability distribution functions. Assume that the means $\mathrm{E}[X_i] = \mu$, and the variances $\mathrm{V}[X_i] = \sigma^2$ are finite. Therefore, for any real x

$$\lim_{N \to \infty} P\left\{ \frac{1}{\sqrt{N}} \sum_{i=1}^{N} \left(\frac{X_i - \mu}{\sigma} \right) \leq x \right\} = P_X(x), \qquad (9.19)$$

9.6 Central Limit Theorem

in which $P_X(x)$ is the cumulative probability function of a standard normal random variable, given by

$$P_X(x) = \frac{1}{\sqrt{2\pi}} \int_{-\infty}^{x} e^{-\frac{x^2}{2}} \, dx. \tag{9.20}$$

This is a result stronger than the Law of Large Numbers, that only states that, for certain conditions, the sample mean converges to the real mean. The Central Limit Theorem effectively provides a probability distribution to which a standard sum converges.

A justification for the result of the theorem can be obtained with the use of the characteristic function, defined as the Fourier transform of the probability density function,

$$P_X(\omega) = \int_{-\infty}^{\infty} p_X(x) e^{-j\omega x} \, dx. \tag{9.21}$$

It is assumed, for simplicity, that the mean of each random variable X_i is zero, and that they are identically distributed. Representing the characteristic function of the variable X_i by $P_{X_i}(\omega)$ and that of the sum of the variables by $P_X(x)$, one obtains

$$P_X(\omega) = \prod_{i=1}^{N} P_{X_i}(\omega). \tag{9.22}$$

The characteristic functions of the variables X_i can be written in terms of their moments of the k-th order,

$$E[X^k] = \frac{1}{(-j)^k} \cdot \frac{\partial^k}{\partial \omega^k} P_X(\omega)\big|_{\omega=0}.$$

Using the expansion in Laurent series,

$$P_{X_i}(\omega) = P_{X_i}(0) + P'_{X_i}(0)\omega + P''_{X_i}(0)\frac{\omega^2}{2!} + \cdots.$$

The expression can be written as a function of the moments,

$$P_{X_i}(\omega) = 1 + (j\omega)E[X_i] + \frac{(j\omega)^2}{2!}E[X_i^2] + \cdots.$$

This implies the expansion of the product 9.22, which becomes

$$P_X(\omega) = 1 + (j\omega)\sum_{i=1}^{N} E[X_i] + \frac{(j\omega)^2}{2!}\sum_{i=1}^{N} E[X_i^2] + \cdots.$$

Considering that the means are zero and that $E[X_i^2] = \sigma_i^2$,

$$P_X(\omega) = 1 + \frac{(j\omega)^2}{2!} \sum_{i=1}^{N} \sigma_i^2 + \cdots .$$

For identically distributed random variables, $\sigma^2 = \sum_{i=1}^{N} \sigma_i^2$, and therefore

$$P_X(\omega) = 1 - \frac{\omega^2}{2}\sigma^2 + \cdots ,$$

which can be approximated as an expansion of the characteristic function for the Gaussian distribution

$$P_X(\omega) = e^{-\frac{\sigma^2 \omega^2}{2}}, \qquad (9.23)$$

and its inverse Fourier transform yields

$$p_X(x) = \frac{1}{\sigma\sqrt{2\pi}} e^{-\frac{x^2}{2\sigma^2}}. \qquad (9.24)$$

In practice, every time an observed variable is the sum of a large number of other variables, that are reasonably well-behaved, then the theorem justifies the assumption that this composition has normal distribution. A classic example is the atmospheric noise, that is composed of countless electromagnetic sources, resulting in the well-known Gaussian noise.

Example: a binary data transmission channel introduces errors with probability P_e. What is the probability that there are n or less errors in a transmission of N bits?

The total number of errors S_N is the sum of independent and identically distributed (i.i.d.) Bernoulli random variables

$$S_N = X_1 + X_2 + \cdots + X_N,$$

the mean value of S_N is $E[S_N] = NP_e$, and the variance is given by $V[S_N] = NP_e(1 - P_e)$.

Applying the Central Limit Theorem, yields

$$P(S_N \leq n) = 1 - P(S_N > n)$$
$$= 1 - P\left(\frac{S_N - NP_e}{\sqrt{NP_e(1-P_e)}} > \frac{n - NP_e}{\sqrt{NP_e(1-P_e)}}\right)$$

$$\approx Q\left(\frac{n - NP_e}{\sqrt{NP_e(1-P_e)}}\right),$$
(9.25)

in which the function $Q(\cdot)$ is given by

$$Q(x) = \frac{1}{\sqrt{2\pi}} \int_x^\infty e^{-t^2/2} dt. \blacksquare$$
(9.26)

9.6.2 Central Limit Theorem for Products

Given N independent and identically distributed positive random variables X_i, with means $E[X_i]$ and variances $V[X_i]$, their product, that is, $X = X_1 X_2 \cdots X_N$ has the following distribution

$$p_X(x) = \frac{1}{x\sigma\sqrt{2\pi}} e^{-\frac{(\ln x - \mu)^2}{2\sigma^2}} u(x),$$
(9.27)

in which

$$\mu = \sum_{i=1}^{N} E[\ln X_i]$$
(9.28)

and

$$\sigma^2 = \sum_{i=1}^{N} V[\ln X_i].$$
(9.29)

To show the theorem, it suffices to introduce the random variable $Y = \ln X = \ln X_1 + \cdots \ln X_N$ and, from the Central Limit Theorem, for a large enough value of N, this variable is approximately normal, with mean μ and variance σ^2. Since $X = e^Y$, it follows that X has a log-normal distribution.

Example: suppose that the random variables have uniform distribution over the interval [0,1] (Papoulis, 1991). Then

$$\mu = \sum_{i=1}^{N} E[\ln X_i] = \sum_{i=1}^{N} \int_0^1 \ln x \, dx = -N,$$

and

$$\sigma^2 = \sum_{i=1}^{N} V[\ln X_i] = \sum_{i=1}^{N} \int_0^1 (\ln x)^2 dx = N.$$

Therefore,

$$p_X(x) = \frac{1}{x\sigma\sqrt{2\pi N}} e^{-\frac{(\ln x + N)^2}{2N}} \mathrm{u}(x).$$

which is the probability density function of a log-normal distribution. ∎

9.7 Pierre-Simon Laplace

Pierre-Simon, marquis de Laplace (1749–1827) was a French mathematician and astronomer whose work was important to the development of several areas of science, including engineering, mathematics, statistics, physics, astronomy, and philosophy. A portrait of Laplace is shown in Figure 9.1.

Laplace formulated the set of equations that bear his name, and developed the Laplace transform, which appears in many branches of mathematics and physics. The Laplacian differential operator is widely used in mathematics. He developed the nebular hypothesis of the origin of the Solar System, and devised the concept of the black hole, when he pointed out that there

Figure 9.1 A portrait of Pierre-Simon, marquis de Laplace. Adapted from: James Posselwhite, www.britannica. com, Public Domain, https://commons.wikimedia.org/w/index.php?curid=11128070

could be massive stars whose gravity is so great that not even light could escape from their surface

Laplace published a paper for the common reader, em 1814, entitled *Essai Philosophique Sur les Probabilités* (A Philosophical Essay on Probability). Later, this work became the introduction to the second edition of the important *Théorie Analytique des Probabilités* (Analytic Theory of Probability), published in 1812.

He applied the theory to the problems of chance and to investigate the causes of phenomena, vital statistics, and future events, and emphasized its importance for physics and astronomy. In this book Laplace included the central limit theorem, and proved that the distribution of errors in large data samples from astronomical observations can be approximated by a Gaussian distribution (Whitrow, 2022).

Laplace became member of the committee of the *Académie des Sciences* (Academy of Sciences) to standardise weights and measures, in 1790, and the committee began working on the metric system advocating a decimal base. But, the work came to an end because of the Reign of Terror established in France, in 1793. During this period the *Académie des Sciences* was closed and some scientists were guillotined. Antoine-Laurent de Lavoisier (1743-1794), a French nobleman and chemist, who was proeminent to the chemical revolution, and who worked with Laplace, was decapitated.

The *Académie des Sciences* was reopened, 1795, as the *Institut National des Sciences et des Arts* (National Institute of Science and Arts). In the same year the *Bureau des Longitudes* (Office of Longitudes) was founded. Joseph-Louis Lagrange (1736–1813), an Italian mathematician and astronomer, and Laplace were among the founding members. Later, Laplace became the leader of the Bureau and the Paris Observatory.

9.8 Problems

1. Prove that convergence in probability, given in the following, implies convergence in distribution.

$$P\{|X_n - X| < \epsilon\} \to 1,$$

when $n \to \infty$.

2. Prove that almost surely convergence, given by the limit

$$P\{\lim_{n \to \infty} X_n = X\} = 1,$$

implies convergence in probability.

3. Show that sure convergence, defined in the following, implies all the other kinds of convergence.
$$\lim_{n\to\infty} X_n(\omega) = X(\omega), \forall \omega \in \Omega,$$
in which Ω is the sample space in which the random variables are defined.

4. Demonstrate that convergence in probability implies convergence in distribution, defined in the following. That is, if $X_n \xrightarrow{P} X$, then $X_n \xrightarrow{d} X$.
$$\lim_{n\to\infty} P_{X_n}(x) = P_X(x),$$
for every number x for which $P_X(x)$ is continuous.

5. Consider that X_n converges in distribution to the random variable X, and prove that for every continuous function $f(x)$, $Y_n = f(X_n)$ converges in distribution to $Y = f(X)$.

6. Consider that X_n converges in distribution to X, and demonstrate that the characteristic function of X_n, $P_{X_n}(\omega)$ converges to the characteristic function of X.

7. Let X_n be a Gaussian random variable, with mean μ_n and variance σ_n^2, and consider that $\mu_n \to \mu$ and $\sigma_n^2 \to \sigma^2$. Prove that X_n converges in distribution to X.

8. Prove that if a sequence of random variables X_n converges almost surely to the random variable X, and if a sequence of random variables Y_n converges almost surely to the random variable Y, then the sum $X_n + Y_n$ converges almost surely to the random variable $Y + X$.

9. Demonstrate that if a sequence of random variables X_n converges almost surely to the random variable X, then X_n converges in probability to the random variable X.

10. Consider that X_n converges in distribution to X. If $f(x)$ is a continuous function, prove that $f(X_n)$ converges in distribution to $f(X)$.

11. Explain the Weak Law of Large Numbers. What are the established conditions for its validity?

A
Formulas and Important Inequalities

"There is no branch of mathematics, however abstract, which may not some day be applied to phenomena of the real world."
Nikolai Lobachevsky

Schwartz Inequality

$$\left| \text{Re} \int_{-\infty}^{\infty} x(t) y^*(t) \mathrm{d}t \right| \leq \left[\int_{-\infty}^{\infty} |x^2(t)| \mathrm{d}t \int_{-\infty}^{\infty} |y^2(t)| \mathrm{d}t \right]^{1/2}$$

the equality verifies for $x(t) = ky(t)$, in which k is a constant.

Holder Inequality

$$\left| \sum_{i=0}^{\infty} a_i b_i \right| \leq \left[\sum_{i=0}^{\infty} |a_i^2| \sum_{i=0}^{\infty} |b_i^2| \right]^{1/2}$$

the equality is verified for $a_i = k b_i$.

Other Inequalities

1. Consider that $w_1, w_2, \ldots, w_N$ are arbitrary positive numbers and let $q_1, q_2, \ldots, q_N$ be positive numbers, such that $\sum_{i=1}^{N} p_i = 1$. Then,

$$\sum_{i=1}^{k} q_i w_i \log \sum_{i=1}^{k} q_i w_i \leq \sum_{i=1}^{k} q_i w_i \log q_i,$$

this equality id verified because of the continuity and convexity of the function $x \log x$ in the closed interval $(0, 1)$. The equality occurs if $q_i = q$, for all $i = 1, \ldots, N$ (Nedoma, 1957).

2. Let $p_1, p_2, \ldots, p_N$ be arbitrary positive numbers and let $q_1, q_2, \ldots, q_N$ be positive numbers, such that their sum is unit, that is, $\sum_{i=1}^{N} p_i = 1$. Then,

$$q_1^{p_1} \cdots q_k^{p_k} \leq \sum_{i=1}^{k} p_i q_i, \quad \sum_{i=1}^{N} q_i = 1, \; q_i \geq 0,$$

the equality occurs for $k = N$ and $q_i = q$ (Ash, 1990).

3. If $p_1 \geq p_2 \geq \cdots \geq p_k$ and $q_1 \geq q_2 \geq \cdots \geq q_k$, then (Chebyshev Inequality)

$$\left(\frac{p_1 + p_2 \cdots p_k}{k}\right)\left(\frac{q_1 + q_2 \cdots q_k}{k}\right) \leq \frac{1}{k}\sum_{i=1}^{k} p_i q_i,$$

or

$$\sum_{i=1}^{k} q_i \sum_{j=1}^{k} p_j \leq k \sum_{i=1}^{k} p_i q_i,$$

the equality is verified only and only if $p_i = p$ or $q_i = q$, for $1 \leq i \leq k$ (Gradshteyn and Ryzhik, 1990).

4. Consider $w_1, w_2, \ldots, w_N$ positive arbitrary numbers and let $p_1, p_2, \ldots, p_N$ be positive numbers, such that $\sum_{i=1}^{N} p_i = 1$. Then,

$$w_1^{p_1} \cdots w_k^{p_k} \leq \sum_{i=1}^{k} p_i w_i, \quad \sum_{i=1}^{N} p_i = 1, \; p_i \geq 0,$$

the equality is obtained for $k = N$ and $w_i = w$. The inequality works even if any p_i equals zero, if $\sum_{i=1}^{N} p_i = 1$ (Ash, 1990).

Final Value Theorem

$$\lim_{t \to \infty} x(t) = \lim_{\omega \to 0} [j\omega X(\omega)]$$

MacLaurin Series

$$f(x) = f(0) + f'(0)x + \frac{1}{2!}f''(0)x^2 + \frac{1}{3!}f'''(0)x^3 + \cdots$$

Bessel Identities

$$J_n(x) = \left(\frac{x}{2}\right)^n \sum_{k=0}^{\infty} \frac{(-x^2/4)^k}{k!(n+k)!}$$

$$= \frac{j^{-n}}{\pi} \int_0^\pi e^{jx\cos\theta} \cos(n\theta) d\theta$$

$$J_n(xe^{jm\pi}) = e^{jnm\pi} J_n(x)$$

$$I_n(x) = j^n J_n(x/j)$$

$$= \frac{1}{\pi} \int_0^\pi e^{x\cos\theta} \cos(n\theta) d\theta$$

Complex Identities

$$z = x + jy = \sqrt{x^2 + y^2} e^{j\tan^{-1}(y/x)}$$
$$z^* = x - jy = \sqrt{x^2 + y^2} e^{-j\tan^{-1}(y/x)}$$
$$|z|^2 = zz^* = x^2 + y^2$$
$$\text{Re}\{z\} = \frac{1}{2}[z + z^*]$$
$$\text{Im}\{z\} = \frac{1}{2j}[z - z^*]$$
$$\text{Re}\{z_1 z_2\} = \text{Re}\{z_1\}\text{Re}\{z_2\} - \text{Im}\{z_1\}\text{Im}\{z_2\}$$
$$\text{Im}\{z_1 z_2\} = \text{Re}\{z_1\}\text{Im}\{z_2\} + \text{Im}\{z_1\}\text{Re}\{z_2\}$$

Trigonometric Identities

$$\text{sen}\,\theta = \frac{1}{2j}(e^{j\theta} - e^{-j\theta})$$

$$\cos\theta = \frac{1}{2}(e^{j\theta} + e^{-j\theta})$$

$$e^{\pm j\theta} = \cos\theta \pm j\text{sen}\,\theta$$

$$\text{sen}\,(\alpha \pm \beta) = \text{sen}\,\alpha \cos\beta \pm \cos\alpha \text{sen}\,\beta$$

$$\cos(\alpha \pm \beta) = \cos\alpha\cos\beta \mp \operatorname{sen}\alpha\operatorname{sen}\beta$$

$$\tan(\alpha \pm \beta) = \frac{\tan\alpha \pm \tan\beta}{1 \mp \tan\alpha\tan\beta}$$

$$\operatorname{sen}\alpha\operatorname{sen}\beta = \frac{1}{2}\cos(\alpha - \beta) - \frac{1}{2}\cos(\alpha + \beta)$$

$$\cos\alpha\cos\beta = \frac{1}{2}\cos(\alpha - \beta) + \frac{1}{2}\cos(\alpha + \beta)$$

$$\operatorname{sen}\alpha\cos\beta = \frac{1}{2}\operatorname{sen}(\alpha - \beta) + \frac{1}{2}\operatorname{sen}(\alpha + \beta)$$

$$e^{\pm j\theta} = \cos\theta \pm j\operatorname{sen}\theta$$

$$\cos\theta = \frac{1}{2}(e^{j\theta} + e^{-j\theta})$$

$$\operatorname{sen}\theta = (e^{j\theta} - e^{-j\theta})/2j$$

$$\operatorname{sen}^2\theta + \cos^2\theta = 1$$

$$\cos^2\theta - \operatorname{sen}^2\theta = \cos 2\theta$$

$$\cos^2\theta = \frac{1}{2}(1 + \cos 2\theta)$$

$$\cos^3\theta = \frac{1}{4}(3\cos\theta + \cos 3\theta)$$

$$\operatorname{sen}^2\theta = \frac{1}{2}(1 - \cos 2\theta)$$

$$\operatorname{sen}^3\theta = \frac{1}{4}(3\operatorname{sen}\theta - \operatorname{sen}3\theta)$$

Series Expansions

$$(1+x)^n = 1 + nx + \frac{n(n-1)}{2!}x^2 + \cdots \quad |nx| < 1$$

$$e^x = 1 + x + \frac{1}{2!}x^2 + \cdots$$

$$a^x = 1 + x\ln a + \frac{1}{2!}(x\ln a)^2 + \cdots$$

$$\ln(1+x) = x - \frac{1}{2}x^2 + \frac{1}{3}x^3 + \cdots$$

$$\operatorname{sen} x = x - \frac{1}{3}!x^3 + \frac{1}{5}!x^5 - \cdots$$

$$\cos x = 1 - \frac{1}{2!}x^2 + \frac{1}{4!}x^4 - \cdots$$

$$\tan x = x + \frac{1}{3}x^3 + \frac{1}{5}x^5 + \cdots$$

$$\sum_{m=0}^{M} x^m = \frac{(x^M - 1)}{(x - 1)}$$

$$e^{a\cos b} = \sum_{i=0}^{\infty} \epsilon_i I_i(a) \cos(ib), \quad \epsilon_0 = 1, \; \epsilon_i = 2, \; i \geq 1$$

$$\cos(x\operatorname{sen}\theta) = J_0(x) + 2\sum_{k=1}^{\infty} J_{2k}(x)\cos(2k\theta)$$

$$\operatorname{sen}(x\operatorname{sen}\theta) = 2\sum_{k=0}^{\infty} J_{2k+1}(x)\operatorname{sen}[(2k+1)\theta]$$

$$\cos(x\cos\theta) = J_0(x) + 2\sum_{k=0}^{\infty} (-1)^k J_{2k}(x)\cos(2k\theta)$$

$$\operatorname{sen}(x\cos\theta) = 2\sum_{k=0}^{\infty} (-1)^k J_{2k+1}(x)\cos[(2k+1)\theta]$$

$$\int_0^\infty \frac{dx}{1+x^n} = \frac{(\pi/n)}{\operatorname{sen}(\pi/n)}, \quad n > 1 \qquad \int_0^\infty \frac{dx}{(a^2+x^2)^2} = \frac{\pi}{4a^3}$$

$$\int_0^\infty \frac{x^u\,dx}{1+x^n} = \left(\frac{\pi}{n}\right)\operatorname{cossec}\left[\frac{(u+1)\pi}{n}\right] \qquad \int_0^\infty \frac{dx}{1+x^2} = \tan^{-1}(b)$$

Indefinite Integrals

$$\int \operatorname{sen} ax\,dx = -\frac{1}{a}\cos ax$$

$$\int \cos ax\,dx = \frac{1}{a}\operatorname{sen} ax$$

$$\int \operatorname{sen}^2 ax\,dx = \frac{x}{2} - \frac{\operatorname{sen} 2ax}{4a}$$

$$\int \cos^2 ax\, dx = \frac{x}{2} + \frac{\operatorname{sen} 2ax}{4a}$$

$$\int \operatorname{sen} ax \cos ax\, dx = \frac{1}{2a}\operatorname{sen}^2(ax)$$

$$\int x\operatorname{sen} ax\, dx = \frac{1}{a^2}(\operatorname{sen} ax - ax\cos ax)$$

$$\int x\cos ax\, dx = \frac{1}{a^2}(\cos ax + ax\operatorname{sen} ax)$$

$$\int x^2\operatorname{sen} ax\, dx = \frac{1}{a^3}(2ax\operatorname{sen} ax + 2\cos ax - a^2 x^2 \cos ax)$$

$$\int x^2 \cos ax\, dx = \frac{1}{a^3}(2ax\cos ax - 2\operatorname{sen} ax + a^2 x^2 \operatorname{sen} ax)$$

$$\int \operatorname{sen} ax \operatorname{sen} bx\, dx = \frac{\operatorname{sen}(a-b)x}{2(a-b)} - \frac{\operatorname{sen}(a+b)x}{2(a+b)} \qquad a^2 \neq b^2$$

$$\int \cos ax \cos bx\, dx = \frac{\operatorname{sen}(a-b)x}{2(a-b)} + \frac{\operatorname{sen}(a+b)x}{2(a+b)}x \qquad a^2 \neq b^2$$

$$\int \operatorname{sen} ax \cos bx\, dx = \frac{\cos(a-b)x}{2(a-b)} - \frac{\cos(a+b)x}{2(a+b)} \qquad a^2 \neq b^2$$

$$\int e^{ax}\, dx = \frac{1}{a}e^{ax}$$

$$\int xe^{ax}\, dx = \frac{1}{a^2}e^{ax}(ax-1)$$

$$\int x^2 e^{ax}\, dx = \frac{1}{a^3}e^{ax}(a^2 x^2 - 2ax + 2)$$

$$\int e^{ax} \operatorname{sen} bx\, dx = \frac{1}{a^2+b^2}e^{ax}(a\operatorname{sen} bx - b\cos bx)$$

$$\int e^{ax} \cos bx\, dx = \frac{1}{a^2+b^2}e^{ax}(a\cos bx + b\operatorname{sen} bx)$$

$$\int \left[\frac{\operatorname{sen} ax}{x}\right]^2 dx = a\int \frac{\operatorname{sen} 2ax}{x}dx - \frac{\operatorname{sen}^2 ax}{x}$$

$$\int \frac{dx}{a^2+b^2 x^2} = \frac{1}{ab}\tan^{-1}\left(\frac{bx}{a}\right)$$

$$\int \frac{x^2 \mathrm{d}x}{a^2 + b^2 x^2} = \frac{x}{b^2} - \frac{a}{b^3} \tan^{-1}\left(\frac{bx}{a}\right)$$

$$\int \frac{\mathrm{d}x}{(a^2 + b^2 x^2)^2} = \frac{x}{2a^2(a^2 + b^2 x^2)} + \frac{1}{2ab^3} \tan^{-1}\left(\frac{bx}{a}\right)$$

$$\int \frac{x^2 \mathrm{d}x}{(a^2 + b^2 x^2)^2} = \frac{-x}{2b^2(a^2 + b^2 x^2)} + \frac{1}{2ab^3} \tan^{-1}\left(\frac{bx}{a}\right)$$

$$\int \frac{\mathrm{d}x}{(a^2 + b^2 x^2)^3} = \frac{x}{4a^2(a^2 + b^2 x^2)^2} + \frac{3x}{8a^4(a^2 + b^2 x^2)} + \frac{3}{8a^5 b} \tan^{-1}\left(\frac{bx}{a}\right)$$

Definite Integrals

$$\int_0^\infty \frac{\operatorname{sen} ax}{x} \mathrm{d}x = \begin{cases} \pi/2 & a > 0 \\ 0 & a = 0 \\ -\pi/2 & a < 0 \end{cases}$$

$$\int_0^x \frac{\operatorname{sen} u}{u} \mathrm{d}u \stackrel{\triangle}{=} \operatorname{Si}(x) \quad \text{the integral is a function of } x$$

$$\int_0^\infty \frac{\operatorname{sen}^2 ax}{x^2} \mathrm{d}x = |a|\pi/2$$

$$\int_0^\infty e^{-ax^2} \mathrm{d}x = \frac{1}{2}\sqrt{\pi/a}$$

$$\int_0^\infty x e^{-ax^2} \mathrm{d}x = \frac{1}{2a}$$

$$\int_0^\infty x^2 e^{-ax^2} \mathrm{d}x = \frac{1}{4a}\sqrt{\pi/a}$$

$$\int_0^\infty \frac{\mathrm{d}x}{(x^2 + a^2)(x^2 + b^2)} = \frac{\pi}{2ab(a+b)} \qquad a > 0, b > 0$$

$$\int_0^\infty \frac{\mathrm{d}x}{ax^4 + b} = \frac{\pi}{2\sqrt{2b}} \left(\frac{b}{a}\right)^{1/4} \qquad ab > 0$$

$$\int_0^\infty \frac{\mathrm{d}x}{ax^6 + b} = \frac{\pi}{3b} \left(\frac{b}{a}\right)^{1/6} \qquad ab > 0$$

Useful Summations

$$\sum_{n=0}^{\infty} \alpha^n = \frac{1}{1-\alpha}, \ |\alpha| < 1$$

$$\sum_{n=k}^{\infty} \alpha^n = \frac{\alpha^k}{1-\alpha}, \ |\alpha| < 1$$

$$\sum_{n=k}^{N} \alpha^n = \frac{\alpha^k - \alpha^{N+1}}{1-\alpha}, \ \alpha \neq 1$$

$$\sum_{n=0}^{N-1} \alpha^n = \begin{cases} \frac{1-\alpha^N}{1-\alpha}, & \text{if } |\alpha| \neq 1 \\ N, & \text{if } \alpha = 1 \end{cases}$$

B

Fourier Transform

"The pure mathematician, like the musician, is a free creator of his world of ordered beauty."
Bertrand Russell

The Fourier Transform and Some Properties

This appendix presents a summary of properties of the Fourier transform. The Parseval theorem is also presented (Hsu, 1973b) (Lathi, 1989).

- Definition: $F(\omega) = \int_{-\infty}^{\infty} f(t)e^{-j\omega t}dt$
- Inverse: $f(t) = \frac{1}{2\pi}\int_{-\infty}^{\infty} F(\omega)e^{j\omega t}d\omega$
- Magnitude and phase: $F(\omega) = |F(\omega)|e^{j\theta(\omega)}$
- Even $f(t)$ function: $F(\omega) = 2\int_{0}^{\infty} f(t)\cos\omega t dt$
- Odd $f(t)$ function: $F(\omega) = -2j\int_{0}^{\infty} f(t)\sin\omega t dt$
- Area under the curve in time: $F(0) = \int_{-\infty}^{\infty} f(t)dt$
- Area under the transform: $f(0) = \frac{1}{2\pi}\int_{-\infty}^{\infty} F(\omega)d\omega$
- Linearity: $\alpha f(t) + \beta g(t) \leftrightarrow \alpha F(\omega) + \beta G(\omega)$
- Parseval's Theorem
    - $\int_{-\infty}^{\infty} f(t)g(t)dt = \frac{1}{2\pi}\int_{-\infty}^{\infty} F(\omega)G^*(\omega)d\omega$
    - $\int_{-\infty}^{\infty} |f(t)|^2 dt = \frac{1}{2\pi}\int_{-\infty}^{\infty} |F(\omega)|^2 d\omega$
    - $\int_{-\infty}^{\infty} f(\omega)G(\omega)d\omega = \int_{-\infty}^{\infty} F(\omega)g(\omega)d\omega$

B.1 Table of Fourier Transforms

This section presents a set of Fourier transforms. Some properties are also presented, which help computing the transforms (Hsu, 1973b), (Spiegel, 1976), (Lathi, 1989), (Gradshteyn and Ryzhik, 1990), (Oberhettinger, 1990).

$f(t)$	$F(\omega)$
$f(at)$	$\frac{1}{|a|}F(\frac{\omega}{a})$
$f(-t)$	$F(-\omega)$
$f^*(t)$	$F^*(-\omega)$
$f(t-\tau)$	$F(\omega)e^{-j\omega\tau}$
$f(t)e^{j\omega_0 t}$	$F(\omega-\omega_0)$
$f(t)\cos\omega_0 t$	$\frac{1}{2}F(\omega-\omega_0)+\frac{1}{2}F(\omega+\omega_0)$
$f(t)\sin\omega_0 t$	$\frac{1}{2j}F(\omega-\omega_0)-\frac{1}{2j}F(\omega+\omega_0)$
$F(t)$	$2\pi f(-\omega)$
$f'(t)$	$j\omega F(\omega)$
$f^{(n)}(t)$	$(j\omega)^n F(\omega)$
$\int_{-\infty}^{t} f(x)dx$	$\frac{1}{j\omega}F(\omega)+\pi F(0)\delta(\omega)$
$-jtf(t)$	$F'(\omega)$
$(-jt)^n f(t)$	$F^{(n)}(\omega)$
$f(t)*g(t)=\int_{-\infty}^{\infty}f(\tau)g(t-\tau)dx$	$F(\omega)G(\omega)$
$\delta(t)$	1
$\delta(t-\tau)$	$e^{-j\omega\tau}$
$\delta'(t)$	$j\omega$
$\delta^{(n)}(t)$	$(j\omega)^n$

B.1 Table of Fourier Transforms

$f(t)$	$F(\omega)$
$f(t)g(t)$	$\frac{1}{2\pi}F(\omega)*G(\omega) = \frac{1}{2\pi}\int_{-\infty}^{\infty}F(\phi)G(\omega-\phi)d\phi$
$e^{-at}u(t)$	$\frac{1}{a+j\omega}$
$e^{-a|t|}$	$\frac{2a}{a^2+\omega^2}$
e^{-at^2}	$\sqrt{\frac{\pi}{a}}e^{-\omega^2/(4a)}$
te^{-at^2}	$j\sqrt{\frac{\pi}{4a^3}}\omega e^{-\omega^2/(4a)}$
$p_T(t) = \begin{cases} 0 & \text{para } |t|>T/2 \\ A & \text{para } |t|\leq T/2 \end{cases}$	$AT\frac{\sin(\frac{\omega T}{2})}{(\frac{\omega T}{2})}$
$\frac{\sin at}{\pi t}$	$p_{2a}(\omega)$
$te^{-at}u(t)$	$\frac{1}{(a+j\omega)^2}$
$\frac{t^{n-1}}{(n-1)!}e^{-at}u(t)$	$\frac{1}{(a+j\omega)^n}$
$e^{-at}\sin bt\, u(t)$	$\frac{b}{(a+j\omega)^2+b^2}$
$e^{-at}\cos bt\, u(t)$	$\frac{a+j\omega}{(a+j\omega)^2+b^2}$
$\frac{1}{a^2+t^2}$	$\frac{\pi}{a}e^{-a|\omega|}$
$\frac{t}{a^2+t^2}$	$j\pi e^{-a|\omega|}[u(-\omega)-u(\omega)]$
$\frac{\cos bt}{a^2+t^2}$	$\frac{\pi}{2a}[e^{-a|\omega-b|}+e^{-a|\omega+b|}]$
$\frac{\sin bt}{a^2+t^2}$	$\frac{\pi}{2aj}[e^{-a|\omega-b|}-e^{-a|\omega+b|}]$
$\sin bt^2$	$\frac{\pi}{2b}\left[\cos\frac{\omega^2}{4b}-\sin\frac{\omega^2}{4b}\right]$
$\cos bt^2$	$\frac{\pi}{2b}\left[\cos\frac{\omega^2}{4b}+\sin\frac{\omega^2}{4b}\right]$
$\text{sech } bt$	$\frac{\pi}{b}\text{sech}\frac{\pi\omega}{2b}$
$\ln\left[\frac{x^2+a^2}{x^2+b^2}\right]$	$\frac{2e^{-b\omega}-2e^{-a\omega}}{\pi\omega}$

$f(t)$	$F(\omega)$
$f_P(t) = \frac{1}{2}[f(t) + f(-t)]$	$\text{Re}(\omega)$
$f_I(t) = \frac{1}{2}[f(t) - f(-t)]$	$j\text{Im}(\omega)$
$f(t) = f_P(t) + f_I(t)$	$F(\omega) = \text{Re}(\omega) + j\text{Im}\omega$
$e^{j\omega_0 t}$	$2\pi\delta(\omega - \omega_0)$
$\cos\omega_0 t$	$\pi[\delta(\omega - \omega_0) + \delta(\omega + \omega_0)]$
$\sin\omega_0 t$	$-j\pi[\delta(\omega - \omega_0) - \delta(\omega + \omega_0)]$
$\sin\omega_0 t u(t)$	$\frac{\omega_0}{\omega_0^2 - \omega^2} + \frac{\pi}{2j}[\delta(\omega - \omega_0) - \delta(\omega + \omega_0)]$
$\cos\omega_0 t u(t)$	$\frac{j\omega}{\omega_0^2 - \omega^2} + \frac{\pi}{2}[\delta(\omega - \omega_0) + \delta(\omega + \omega_0)]$
$u(t)$	$\pi\delta(\omega) + \frac{1}{j\omega}$
$u(t - \tau)$	$\pi\delta(\omega) + \frac{1}{j\omega}e^{-j\omega\tau}$
$tu(t)$	$j\pi\delta'(\omega) - \frac{1}{\omega^2}$
1	$2\pi\delta(\omega)$
t	$2\pi j\delta'(\omega)$
t^n	$2\pi j^n \delta^{(n)}(\omega)$
$|t|$	$\frac{-2}{\omega^2}$
$\frac{1}{t}$	$\pi j - 2\pi j u(\omega)$
$\frac{1}{t^n}$	$\frac{(-j\omega)^{n-1}}{(n-1)!}[\pi j - 2\pi j u(\omega)]$
$u(t) - u(-t)$	$\frac{2}{j\omega}$
$\frac{1}{e^{2t} - 1}$	$\frac{-j\pi}{2}\coth\frac{\pi\omega}{2} + \frac{j}{\omega}$
$\delta_T(t) = \sum_{n=-\infty}^{\infty} \delta(t - nT)$	$\omega_0 \delta_{\omega_0}(\omega) = \omega_0 \sum_{n=-\infty}^{\infty} \delta(\omega - n\omega_0)$

B.1 Table of Fourier Transforms

$f(t)$	$F(\omega)$
$\cos\left(\frac{t^2}{4a} - \frac{1}{4}\pi\right)$	$2\sqrt{\pi a}\cos(a\omega^2)$
$\sin\left(\frac{t^2}{4a} + \frac{1}{4}\pi\right)$	$2\sqrt{\pi a}\sin(a\omega^2)$
$\frac{\Gamma(1-s)\sin\left(\frac{1}{2}s\pi\right)}{|t|^{1-s}}$	$\pi|\omega|^{-s}\quad 0<\mathrm{Re}(s)<1$
$\frac{1}{\sqrt{\frac{\sqrt{\alpha^2+t^2}+\alpha}{\sqrt{\alpha^2+t^2}}}}$	$\frac{\sqrt{2\pi}}{|\omega|}$ $\sqrt{\frac{2\pi}{|\omega|}}e^{-\alpha|\omega|}$
$\frac{\cos(\frac{1}{2}\alpha)\cosh(\frac{1}{2}t)}{\cosh(t)+\cos(\alpha)}$	$\frac{\pi\cosh(\alpha\omega)}{\cosh(\pi\omega)}\quad -\pi<\alpha<\pi$
$\frac{\sin(\alpha)}{\cosh(t)+\cos(\alpha)}$	$\frac{\pi\sinh(\alpha\omega)}{\sinh(\pi\omega)}\quad -\pi<\alpha<\pi$
$J_0(\alpha t)$	$\frac{2}{\sqrt{\alpha^2-\omega^2}}\quad |\omega|<\alpha$
	$0\quad |\omega|>\alpha$
$J_0(\alpha\sqrt{b^2-t^2})\,|t|<b$	$2\frac{\sin[b(\alpha^2+\omega^2)^{\frac{1}{2}}]}{\sqrt{\alpha^2+\omega^2}}$
0	$|t|>b$
$j^n J_{n+\frac{1}{2}}(t)$	$\sqrt{2}P_n(\omega)\,|\omega|<1$
	$0\quad |\omega|>1$
$J_0(\alpha\sqrt{t^2+b^2})$	$\frac{2\cos(b\sqrt{\alpha^2-\omega^2})}{\sqrt{\alpha^2-\omega^2}}\quad |\omega|<\alpha$
	$0\quad |\omega|>\alpha$
$J_0(\alpha\sqrt{t^2-b^2})$	$\frac{2\cosh(b\sqrt{\alpha^2-\omega^2})}{\sqrt{\alpha^2-\omega^2}}\quad |\omega|<\alpha$
	$0\quad |\omega|>a$

$f(t)$	$F(\omega)$
$\dfrac{1}{(\alpha-jt)^\nu}$	$2\pi\omega^{\nu-1}e^{-\alpha\omega}/\Gamma(\nu), \quad \omega>0$
$\operatorname{Re}\alpha>0, \quad \operatorname{Re}\nu>0$	$0, \qquad\qquad\qquad\qquad \omega<0$
$\dfrac{1}{(\alpha+jt)^\nu}$	$-2\pi(-\omega)^{\nu-1}e^{\alpha\omega}/\Gamma(\nu), \quad \omega<0$
$\operatorname{Re}\nu>0, \quad \operatorname{Re}\alpha>0$	$0, \qquad\qquad\qquad\qquad \omega>0$
$\dfrac{1}{(t^2+\alpha^2)(jt)^\nu}$	$\pi\alpha^{-\nu-1}e^{-|\omega|\alpha}$
$|\nu|<1, \quad \operatorname{Re}\alpha>0$	$\arg(jt)=1/2\pi, (t>0)$
	$\arg(jt)=-1/2\pi, (t<0)$
$\dfrac{1}{(t^2+\alpha^2)(\beta+jt)^\nu} \quad \operatorname{Re}\nu>-1$	$\dfrac{\pi e^{-\alpha\omega}}{\alpha(\alpha+\beta)^\nu}, \quad \omega>0$
	$\operatorname{Re}\alpha>0, \quad \operatorname{Re}\beta>0$
$\dfrac{1}{(x^2+\alpha^2)(\beta-jt)^\nu}$	$\dfrac{\pi(\beta-\alpha)^\nu e^{\alpha\omega}}{\alpha}, \quad \omega>0$
$\operatorname{Re}\nu>-1, \quad \operatorname{Re}\alpha>0$	$\operatorname{Re}\beta>0, \quad \alpha\neq\beta$
$\dfrac{1}{(\alpha-e^{-t})e^{\lambda t}}$	$\pi\alpha^{\lambda-1+j\omega}\cot(\pi\lambda+j\pi\omega)$
	$0<\operatorname{Re}\lambda<1, \quad \alpha>0$
$\dfrac{1}{(\alpha+e^{-t})e^{\lambda t}}$	$\pi\alpha^{\lambda-1+j\omega}\operatorname{cossec}(\pi\lambda+j\pi\omega)$
	$0<\operatorname{Re}\lambda<1, \quad -\pi<\arg\alpha<\pi$
$\dfrac{t}{(\alpha+e^{-t})e^{\lambda t}}$	$\pi\alpha^{\lambda-1+j\omega}\operatorname{cossec}(\pi\lambda+j\pi\omega)$
$0<\operatorname{Re}\lambda<1, \quad -\pi<\arg\alpha<\pi$	$\times[\log\alpha-\pi\cot(\pi\lambda+j\pi\omega)]$

B.1 Table of Fourier Transforms

$f(t)$	$F(\omega)$
$\dfrac{t^2}{(1+e^{-t})e^{\lambda t}}$ $0 < \operatorname{Re}\lambda < 1$	$\pi^3 \operatorname{cossec}^3(\pi\lambda + j\omega\pi)[2 - \sin^2(\pi\lambda + j\omega\pi)]$
$\dfrac{1}{(\alpha+e^{-t})(\beta+e^{-t})e^{\lambda t}}$ $0 < \operatorname{Re}\lambda < 2, \quad \beta \neq \alpha$ $|\arg\alpha| < \pi, \quad |\arg\beta| < \pi$	$\pi(\beta-\alpha)^{-1}(\alpha^{\lambda-1+j\omega} - \beta^{\lambda-1+j\omega})$ $\times \operatorname{cossec}(\pi\lambda + j\omega\pi)$
$\dfrac{t}{(\alpha+e^{-t})(\beta+e^{-t})e^{\lambda t}}$ $0 < \operatorname{Re}\lambda < 2, \quad \alpha \neq \beta$ $|\arg\alpha| < \pi, \quad |\arg\beta| < \pi$	$\dfrac{\pi(\alpha^{\lambda-1+j\omega}\log\alpha - \beta^{\lambda-1+j\omega}\log\beta)}{(\alpha-\beta)\sin(\lambda\pi+j\omega\pi)}$ $+ \dfrac{\pi^2(\alpha^{\lambda-1+j\omega} - \beta^{\lambda-1+j\omega})\cos(\lambda\pi+j\omega\pi)}{(\beta-\alpha)\sin^2(\lambda\pi+j\omega\pi)}$
$\dfrac{1}{(1+e^{-t})^n e^{\lambda t}}$ $n = 1, 2, 3, \cdots, \quad 0 < \operatorname{Re}\alpha < n$	$\pi \operatorname{cossec}(\pi\lambda + j\omega\pi) \prod_{j=1}^{n-1}(j - \lambda - j\omega)/(n-1)!$
$e^{-\lambda t}\log|1 - e^{-t}|$ $-1 < \operatorname{Re}\lambda < 0$	$\pi(\lambda + j\omega)^{-1}\cot(\pi\lambda + j\omega\pi)$
$e^{-\lambda t}\log(1 + e^{-t})$ $-1 < \operatorname{Re}\lambda < 0$	$\pi(\lambda + j\omega)^{-1}\operatorname{cossec}(\pi\lambda + j\omega\pi)$
$e^{-\lambda t}\log\left(\dfrac{|1+e^{-t}|}{|1-e^{-t}|}\right)$ $|\operatorname{Re}\lambda| < 1$	$\pi(\lambda + j\omega)^{-1}\tan(\tfrac{1}{2}\pi\lambda + \tfrac{1}{2}j\omega\pi)$
$\dfrac{1}{(\sinh t + \sinh\alpha)} \quad \alpha > 0$	$-\pi j e^{j\alpha\omega}\operatorname{sech}\alpha \operatorname{cossech}(\pi\omega) \times [\cosh(\pi\omega) - e^{-2j\alpha\omega}]$
$\dfrac{1}{[\Gamma(\nu-t)\Gamma(\mu+t)]}$	$[2\cos(\tfrac{1}{2}\omega)]^{\mu+\nu-2} e^{\tfrac{1}{2}j\omega(\mu-\nu)} \times [\Gamma(\mu+\nu-1)]^{-1}, \quad |\omega| < \pi$ $0, \quad |\omega| > \pi$

C
Commented Bibliography

"A man has died and his body turned into sand.
All his relatives disintegrated into dust.
It is by his writings that he will be remembered."
Inscription made by an Egyptian scribe, four thousand years ago (Fischer, 2007).

Books on Sets, Measure and Probability

The following list presents short comments on books that deal with subjects related to the Theory of Advanced Sets, Theory of Measure and Probability Theory, as well as related topics. Some of them are reprints and others may be out of print. Most of the books in the list are cited as references by other authors.

1. Aczél, J. and Daroczy, Z., On Measures of Information and their Characterization. Academic Press, New York (1975).

   This book presents several definitions of entropy with the appropriate mathematical rigor the topic requires, being considered a reference on the subject.

2. Adams, M. and Guillemin, V., Measure Theory and Probability. BirkHäuser, Boston, USA (1996).

   This book uses probability theory to introduce measure theory at the graduate level. Lebesgue and Riemann integrals are discussed and compared in a formal and clear way. It also features a chapter on Fourier analysis.

3. Ang, A. H.-S. and Tang, W. H., Probability Concepts in Engineering and Design. John Wiley & Sons, New York, USA (1975).

Written for a first probability course on engineering and statistics, it only requires elementary calculus. It contains a chapter on basic theory of sets, but does not discuss measure theory or integration. Brings many applications related to systems planning and design.

4. Ash, R. B., Information Theory. Dover Publications, Inc., New York, USA (1990).

   The book presents the results of Information Theory in a rigorous way, and contains the proof of why the logarithm function is necessary in the definition of entropy.

5. Ash, R. B., Real Analysis and Probability. Academic Press, Inc., Boston, USA (1972).

   It is recommended for a postgraduate course in probability, for the area of Mathematics. It introduces the subject to students with a strong background in general topology. It deals with real analysis, functional analysis, and measure and integration theory.

6. Bartle, R. G., The Elements of Integration. John Wiley & Sons, Inc., New York, USA (1966).

   The book presents the modern theory of integration, which was pioneered by Lebesgue, in 1902. It requires familiarity with real analysis and advanced calculus. The author considers the theory of sets, measure theory and topology are irrelevant, but assumes that the convergence theorems are essential to establish the concept of integral.

7. Beichelt, F., Stochastic Prozesse fur Ingenieure. B. G. Teubner, Stuttgart, Germany (1997).

   It deals with stochastic processes and was written for undergraduate students in engineering, with an interest in communications and systems theory. It features a minimum of set theory and does not address measure theory.

8. Billingsley, P., Probability and Measure. Wiley Series in Probability and Mathematical Statistics. Third Edition. John Wiley & Sons, Inc. (1995).

   The book addresses the idea of measure and its application to probability, using one of them to motivate the other, alternately, in the chapters. The Lebesgue integral is introduced, as well as its relationship with the Lebesgue measure. The convergence of distributions is discussed and some stochastic processes are presented.

9. Blake, I. F., An Introduction to Applied Probability. Robert E. Krieger Publishing Co., Malabar, Florida (1987).

    Highlights probability theory and applications for engineering students, science and administration. The book covers many aspects of the theory, comprehensively, with many examples, but only briefly discusses sets and does not deal with measure theory. There is a chapter on entropy and its applications.

10. Braumann, P. B. T., Teoria da Medida e da Probabilidade – Parte 1: Álgebra dos Conjuntos. Fundação Calouste Gulbenkian, Lisboa, Portugal (1987).

    The book covers measure theory and probability. It was written for students with a strong mathematical background.

11. Bressoud, D. M., A Radical Approachto Lebesgue's Theory of Integration. Cambridge University Press (2008).

    The text is an introduction to Lebesgue's theory of measure and integration, which explores the reader knowledge on the Riemann integral, explores the geometry of the set of reals numbers, presents a historical approach to the development of measure theory and discusses applications to Fourier theory and its justification.

12. Brzezniak, Z. and Zastawniak, T., Basic Stochastic Processes. Springer, Great Britain (1999).

    The book was designed to be taught as a final graduate course in stochastic processes, requiring knowledge of probability theory and calculus. The book follows a singular order in the presentation of subjects, starting with martingales, to then explain Markov and continuous-time processes. Has greater appeal to students of Mathematics.

13. Capiński, M. and Kopp, E., Measure, Integral and Probability. Springer, New York, USA (2005).

    The central concepts of this book are measure and Lebesgue integral. Probability theory was chosen as an application of measure and integral, but the text does not deal with elementary probability.

14. Carter, M. and van Brunt, B., The Lebesgue-Stieltjes Integral – A Practical Introduction. Springer-Verlag, USA (2000).

The book discusses the substitution of the Riemann integral for the Lebesgue integral, as well as other integration theories, with the aim of introducing the Lebesgue-Stieltjes integral in the set of real numbers in a natural way, as an extension of the Riemann integral. Integral and differential calculus is introduced, including double integrals, and there are chapters on Lebesgue and Hilbert spaces.

15. Doob, J. L., Stochastic Processes. John Wiley & Sons, New York, USA (1953).

    This is a classic book on stochastic processes. Probability theory is kept to a minimum. Set theory, measure and related material are presented in an appendix. It targets graduate students with a deep interest in Mathematics.

16. Drake, A. W., Fundamentals of Applied Probability Theory. McGraw-Hill Book Company, New York, USA (1967).

    This is a first book on applied probability theory, intended to present a Physics-based introduction to the subject for undergraduate students. It introduces the basics of set theory, but does not mention measure theory. Its approach is intuitive, with many applications and practice of the basic ideas.

17. Feller, W., An Introduction to Probability Theory and its Applications. John Wiley & Sons, Inc., New York (1968).

    It is an encyclopedia of basic probability, including many examples, distributions and calculations. It is a reference book in the area.

18. Gallager, R. G., Discrete Stochastic Processes. Kluwer Academic Publishers, Boston, USA (1996).

    Presents the subject of stochastic processes, at the undergraduate level, without measure theory, and some mathematical analysis. There are several examples to fix the theory and increase the students' intuitive insight into physical phenomena.

19. Goldberg, S., Probability, An Introduction. Dover Publications, Inc., New York, USA (1960).

    It targets students with a limited mathematical background, being directed towards a bigger audience. It introduces the basics of set algebra but does not mention measure theory.

20. Gray, R. M. and Davisson, L. D., An Introduction to Statistical Signal Processing. Cambridge University Press, Cambridge, United Kingdom (2004).

It was written for undergraduates and aims to introduce the fundamental ideas and the mechanics of random processes for engineers in a way that reflects the underlying mathematics, but does not require a deep background in the area. There is an appendix on set theory and another on integrals, which presents the Lebesgue integral. The book analyzes a large set of stochastic processes.

21. Gray, R. M., Probability, Random Processes, and Ergodic Properties. Springer-Verlag, New York, USA (1988).

The book is more directed to ergodic properties and theorems, with an introduction to measure theory, and is aimed at graduate students in engineering with an inclination towards Mathematics.

22. Gray, R. M. and Davisson, L. D., Random Processes, A Mathematical Approach to Engineers. Prentice-Hall, Inc., Englewood Cliffs, York, USA (1988).

The book is suitable for postgraduate studies. It presents the development of the basic theory and applications of random processes using the language and point of view of rigorous mathematicians on the subject, but requires only a background in Fourier transform theory, linear systems and elementary probability. The tests are simple and provide the basis for the simple justifications of more general and complicated results, which are semi-rigorous. The authors' primary goal is to use intuitive arguments and allow students to read and follow the modern literature in the area.

23. Grimmett, G. R. and Stirzaker, D. R., Probability and Random Processes. Oxford University Press, New York, USA (2001).

Provides an extensive introduction to probability and stochastic processes. It targets undergraduate as well as graduate students. The amount of measure theory is limited and the basics of set theory are presented in an appendix.

24. Gödel, K., O Teorema de Gödel e a Hipótese do Contínuo. Fundação Calouste Gulbenkian, Lisboa, Portugal (1979).

The book written by Kurt Gödel himself, which demonstrated that no mathematical system was complete, presents his fundamental theorem and the continuum hypothesis, which forms the basis for understanding the limits of any axiomatic theory.

25. Gubner, J. A., Probability and Random Processes for Electrical and Computer Engineers. Cambridge University Press, Cambridge, United Kingdom (2006).

    It was written for a postgraduate course in electrical engineering and computing. The book is comprehensive and presents many examples and theory applications, as well as lines of code in Matlab to help students to obtain computational results. Some pages are dedicated to the presentation of theory of sets, but measure theory is not mentioned.

26. Halmos, P. R., Naive Set Theory. D. Van Nostrand Company, Inc., Princeton, USA (1960).

    This book presents set theory with a minimum of philosophical discourse and logical formalism. Axioms related to set theory are introduced in an informal but accurate manner, with very few displayed theorems.

27. Hausdorff, F., Set Theory. Chelsea Publishing Company, New York (1957). Republished by AMS-Chelsea 2005.

    This fundamental book was first published in 1914, in Geman, as *Grundzüge der Mengenlehre*, and was the first comprehensive introduction to set theory. In 1927, Hausdorff published a revised second edition entitled just *Mengenlehre*. The German edition was translated into English under the title Set Theory. Hausdorff described the foundations of mathematics as a tree, with set theory as its trunk and all of the diverse domains of Mathematics as individual branches. This beautiful book established the baiscs for modern mathematics and influenced several subsequent publications.

28. James, B. R., Probabilidade: Um Curso em Nível Intermediário. Projeto Euclides. Instituto de Matemática Pura e Aplicada do CNPq (Impa). Livros Técnicos e Científicos Editora S. A., Rio de Janeiro, Brasil (1981).

    The book deals with probability at a level that lies between the elementary introduction to the subject and a more advanced course based on measure theory and integration. It is generally suitable for the beginning of graduate studies, and requires familiarity with notions of sets and functions, as well as differential and integral calculus. Mathematical expectation is treated with the Stieltjes integral, and the Lebesgue integral is only mentioned. There is a chapter on laws of large numbers and another on the Central Limit Theorem.

29. Kolmogorov, A. N. and Fomin, S. V., Introductory Real Analysis. Dover Publications, Inc., New York, USA (1970).

    This is a classic book, written by the founder of modern Probability Theory. He presents set theory, measurement and integration in a formal and clear way, in the context of functional analysis.

30. Lapin, L. L., Probability and Statistics for Modern Engineering. PWS Engineering, Boston, USA (1983).

    The book is more related to statistics than probability theory, and the theory of sets is slightly presented. Contains an exhaustive collection of probability distributions with their main properties, which is useful for students.

31. Larsen, R. J. and Marx, M. L., An Introduction to Probability and Its Applications. Prentice-Hall, Inc., Englewood Cliffs, York, USA (1985).

    It presents an introductory mathematical treatment of probability, mainly based on examples and case studies. Includes historical information on the subject. The book has a very good chapter on limit theorems, full of historical annotations. The theory of sets is kept to a minimum and it does not deal with measure theory.

32. Leon-Garcia, A., Probability and Random Processes for Electrical Engineering . Addison-Wesley Publishing Co., Reading, Massachusetts (1989).

    The book presents the standard approach of probability and random processes for the electrical engineering course. Numerous examples are used to demonstrate analytic techniques to solve problems. It does not deal with advanced set theory or with measure theory.

33. Lévine, B., Fondements Théoriques de la Radiotechnique Statistique. Éditions de Moscou, Moscow, U.S.S.R (1973).

    This book collection is encyclopedic and presents most of the process theory and statistical communication in an intuitive and applied way.

34. Lipschutz, S., Set Theory and Related Topics. Schaum´s Outline Series. McGraw-Hill, New York, USA (1998).

   This book is a best seller and presents, in a nutshell, the theory of sets and related topics, including hundreds of examples and exercises with answers. It is a good reference book for undergraduate students.

35. Meyer, P., Probabilidade – Aplicações à Estatística. Editora LTC, São Paulo, Brasil (2013).

   It is a textbook for probability and statistics that are used in engineering courses. Some concepts are introduced informally, but definitions and theorems are stated rigorously. The author comments most theorems and definitions.

36. MacLane, S. and Birkhoff, G., Algebra. The Macmillan Company, New York, USA (1968).

   Brings algebra concepts to undergraduate students in a comprehensive way, covering sets, functions, groups, rings, modules, spaces vectors matrices, tensors, affine projective spaces, lattice categories and multilinear algebra.

37. Magalhães, M. N., Probabilidade e Variáveis Aleatórias. Editora da Universidade de São Paulo (Edusp), São Paulo, Brasil (2004).

   It is aimed at postgraduate students at the beginning of the course in the areas of exact sciences, mainly Mathematics and Statistics. The basics of set theory and algebra are presented. The Lebesgue-Stieltjes integral is introduced in connection with the definition of expected value. There are several examples and many proposed exercises in the text, and the answers can be found in the appendix. The book contains the necessary theory on the subject and can be used as a reference.

38. Marques, M. S. F., Teoria da Medida. Editora da Unicamp, Campinas, Brasil (2009).

   The book approaches measure theory from the theory of sets and functions, and introduces some usual measures, such as the Lebesgue-Stieltjes, the Lebesgue, the Wiener and the of Gauss, in addition to notions of integral, product space, signed measure and convergence modes. The book aims to prepare students for advanced studies of probability, statistics and stochastic calculus.

39. Mitzenmacher, M. and Upfal, E., Probability and Computing. Cambridge University Press, New York, USA (2005).

   Indicated for students of computer science and applied mathematics, at the end of the course or at the beginning of postgraduate studies. It presents the probability techniques and examples used to develop probabilistic models and algorithms. Introduction to sets is minimal and measure theory is omitted.

40. Papoulis, A., Probability, Random Variables, and Stochastic Processes. McGraw-Hill, Tokyo (1991).

   The first edition of this book was published in 1965. It has become a classic in the field. and has had several new editions since then. The issues of probability and Stochastic Processes are presented as a deductive discipline, avoiding sophisticated mathematics. Elementary properties of sets are presented, but measure theory is mentioned just *in passant*.

41. Parzen, E., Modern Probability Theory and its Applications. John Wiley & Sons, New York, USA (1960).

   Presents a basic probability course with a brief introduction to space sampling and set theory. It does not require advanced Mathematics, but it does deal with the Poisson processes in detail. The book also presents some historical notes on the literature concerning the Theory of Probabilities.

42. Peebles, Peyton Z., Probability, Random Variables, and Random Signal Principles. McGraw-Hill Book Company, New York, USA (1987).

   Designed for electrical engineering students, the book contains many problems and examples, as well as a useful collection of probability density functions and their properties in one of the appendices. Set theory is minimal and measure theory is not present in the book.

43. Petrov, V. V. e Mordecki, E., Teoría de la Probabilidad. Dirac – Faculdade de Ciências, Universidade de la República, Montevideo, Uruguay (2008).

   It is an initial probability course and has as a requirement only differential and integral calculus. Uses a simple approach to set theory, without reference to measure theory.

44. Phillips, E. R., An Introduction to Analysis and Integration Theory. Dover Publications, Inc., New York, USA (1984).

The book presents an interesting historical introduction to the subject of integration. The book is aimed at students with a high interest in Mathematics. The set theory is presented and the Lebesgue integral is preceded by the study of the real number system, metric and normalized spaces, compression and continuity.

45. Rényi, A., Probability Theory. Dover Publications, Inc., New York, USA (2007).

Alfréd Rényi's book is a classic on probability theory, whose publication could not be seen by the author, who died before the final revision of the text. He contains all the basics of probability, starting with event algebra, going through from the general theory of random variables, to the laws of large numbers and the limiting theorems of probability theory. It also includes an appendix on the theory of information.

46. Ricciardi, Luigi M., Lectures in Applied Mathematics and Informatics. Manchester University Press, Manchester, UK (1989).

This specialized book brings a collection of articles on several areas of mathematics and computer science, plus a useful chapter on generalized entropies.

47. Rosenthal, Jeffrey S., A First Look at Rogorous Probability Theory. World Scientific Publishing Co. Pte. Ltd., Singapore, (2000).

It presents a probability approach based on measure theory, with a discussion of the Lebesgue integral. It also addresses some stochastic processes and Markov chains, in addition to convergence criteria and probability laws.

48. Ross, S. M., Introduction to Probability Models. Academic Press, Inc., Boston, USA (1989).

It is an introduction to elementary probability theory and stochastic processes, suitable for the study of phenomena in engineering, management, operations research, physical and social sciences. The approach is heuristic rather than rigorous, with a minimum of set theory, but complete with many applications.

49. Shilov, G. E. and Gurevich, B. L., Integral, Measure and Derivative: A Unified Approach. Dover Publications, Inc., New York, USA (1977).

The book presents the Riemann, Lebesgue, Stieltjes and Lebesgue-Stieltjes integrals, as well as a full discussion of measurable sets and general measure theory. The book is aimed at postgraduate or senior students in Mathematics.

50. Stoll, R. R., Set Theory and Logic. Dover Publications, Inc., New York, USA (1963).

    Introduces advanced undergraduate students to the fundamentals of Mathematics. Set theory is presented in connection with number systems, and cardinality is placed in the same context.

51. Taylor, A. E., General Theory of Functions and Integration. Dover Publications, Inc., New York, USA (1985).

    Provides a transition from the classical theory of points in the Euclidean space, set theory and the theory of functions, to the abstract ideas of Topology, continuous functions and integration. It introduces measure theory and the Lebesgue integral.

52. Tortrat, A., Calcul des Probabilités et Introduction aux Processus Aléatoires. Masson et Cie Editeurs, Paris, France (1971).

    It is dedicated to graduate students in Mathematics and presents a chapter on measure and integration. Advanced set theory is kept to a minimum. The notation used in the book, in certain passages, is unusual.

53. Ventsel, H., Théorie des Probabilités. Éditions Mir, Moscow (1973).

    Directed to undergraduate students, it requires only basic Mathematics. The focus of the presentation is intuitive, based on frequency relative and applications. It has one chapter on entropy and information theory.

54. Whittle, P., Probability. John Wiley & Sons, London, UK (1970).

    Its aim is to be a first text in probability theory and applications. The book novelty is the axiomatization of the concept of expectation rather than the probability measure. Set theory and measure are kept to a minimum so the book is accessible to engineers and technologists.

55. Wilbur B. Davenport, Jr. and Root, W. L., An Introduction to the Theory of Random Signals and Noise. IEEE Press, New York, USA (1987).

    An appropriate book for a discipline studying statistical noise theory and modulation at the beginning of graduate studies. It presents a brief introduction to the theory of sets and probability, but there is nothing about measure theory. The remaining of the book deals with application of stochastic processes to communications theory and related areas. It is a reference book.

56. Wilcox, H. J. and Myers, D. L., An Introduction to Lebesgue Integration and Fourier Series. Dover Publications, Inc., New York, USA (1994).

    Introduces Riemann and Lebesgue integrals and an introduction to measurable sets and their properties. It also deals with the calculation of Fourier series. It is a suitable book for advanced undergraduate students or graduate students in Mathematics.

57. Williams, R., Probability, Statistics and Random Processes for Engineers. Thomson, Pacific Grove, USA (2003).

    Written for undergraduate students in electrical and computer engineering. It discusses methods of probability, statistics, and random processes, and includes random variables, random processes, statistical inference and confidence, random countable events and reliability. It contains only the information absolutely necessary for the engineering-related topics, but does not cover set and measure theory.

Bibliography

Adams, M. and Guillemin, V., *Measure Theory and Probability*. Birkhäuser Boston, York, USA (1996).

Alencar, M. S., O Matemático que Sabia Medir – I. Artigo para jornal eletrônico na Internet, Jornal do Commercio *On Line*, Recife, Brasil (2008a).

Alencar, M. S., O Matemático que Sabia Medir – II. Artigo para jornal eletrônico na Internet, Jornal do Commercio *On Line*, Recife, Brasil (2008b).

Alencar, M. S., O Matemático que Sabia Medir – III. Artigo para jornal eletrônico na Internet, Jornal do Commercio *On Line*, Recife, Brasil (2008c).

Alencar, M. S., O Matemático que Sabia Medir – IV. Artigo para jornal eletrônico na Internet, Jornal do Commercio *On Line*, Recife, Brasil (2008d).

Alencar, M. S., Probabilidade, Complexidade e Informação – I. Artigo para jornal eletrônico na Internet, Jornal do Commercio *On Line*, Recife, Brasil (2008e).

Alencar, M. S., Probabilidade, Complexidade e Informação – II. Artigo para jornal eletrônico na Internet, Jornal do Commercio *On Line*, Recife, Brasil (2008f).

Alencar, M. S., *Telefonia Digital, Quinta Edição*. Editora Érica Ltda., ISBN 978-85-365-0312-7, São Paulo, Brasil (2011).

Alencar, M. S., Personagens e Fórmulas Desiguais. Artigo para jornal eletrônico na Internet, NE10 – Sistema Jornal do Commercio de Comunicação, Recife, Brasil (2013).

Ash, R. B., *Real Analysis and Probability*. Academic Press, Inc., San Diego, USA (1972).

Ash, R. B., *Information Theory*. Dover Publications, Inc., New York, USA (1990).

Bagaria, J., Set theory. In Zalta, E. N., editor, *The Stanford Encyclopedia of Philosophy*. Winter 2014 edition (2014).

Bayes, T., An essay towards solving a problem in the doctrine of chance. *Philosophical Transactions of the Royal Society of London*, (53):370–418 (1763).

Billingsley, P., *Probability and Measure*. John Wiley & Sons, Hoboken, USA (1995).

Blake, I. F., *An Introduction to Applied Probability*. Robert E. Krieger Publishing Co., Malabar, Florida (1987).

Boyer, C., *História da Matemática*. Editora Edgard Blucher Ltda., São Paulo, Brasil (1974).

Braga, F., Lopes, W., and Alencar, M. S., Cognitive Vehicular Networks: An Overview. *Journal of Procedia Computer Science (ISSN 1877-0509)*, 65(1):Electronic publication (2015).

Braumann, P. B. T., *Teoria da Medida e da Probabilidade – Parte 1: Álgebra dos Conjuntos*. Fundação Calouste Gulbenkian, Lisboa, Portugal (1987).

Bressoud, D. M., *A Radical Approach to Lebesgue's Theory of Integration*. Cambridge University Press, New York, USA (2008).

Britannica, The Editors of Encyclopaedia, *Function*. Encyclopaedia Britannica, www.britannica.com/science/function-mathematics, Acessed 6 May 2022 (2021).

Bronstein, I. and Semendiaev, K., *Manual de Matemática*. Editora Mir, Moscou (1979).

Cantor, G. F. L. P., "Über eine Eigenschaft des Inbergriffes aller reellen algebraishen Zahlen". *Journal für die reine und angewandte Mathematik*, 77:258–262 (1874).

Capiński, M. and Kopp, P. E., *Measure, Integral and Probability*. Springer-Verlag London Ltd., New York, USA (2005).

Cordeiro, J. E., Alencar, M. S., Yashina, M. V., and Tatashev, A. G., Effect of Epidemic Interference on the Performance of M-ASK, M-PSK and M-QAM Modulation Schemes. *Journal of Communications Software and Systems (JCOMSS)*, 17(4):358–365 (2021).

de Laplace, P.-S., *Théorie Analytique des Probabilités*. Courcier, Imprimeur – Librairie pour les Mathématiques et la Marine, Paris, France (1814).

Dunham, W., *Journey through Genius – The Great Theorems of Mathematics*. Penguin Books, New York, USA (1990).

Dunham, W., *The Calculus Gallery – Masterpieces from Newton to Lebesgue*. Princeton University Press, Princeton, USA (2005).

Ericsson, J., Ollila, E., and Koivunen, V., "Statistics for Complex Random Variables Revisited". In *IEEE International Conference on Acoustics, Speech, and Signal Processing (ICASSP 2009)*, pp. 3565–3568, Taipei, Taiwan (2009).

Fatou, P. J. L., Séries trigonométriques et séries de taylor. *Acta Mathematica*, (30):335–42008 (1906).

Fischer, S. R., *História da Escrita*. Editora UNESP, São Paulo, Brasil (2007).

Gabel, R. A. and Roberts, R. A., *Signals and Linear Systems*. John Wiley & Sons, New York, USA (1973).

Gradshteyn, I. S. and Ryzhik, I. M., *Table of Integrals, Series, and Products*. Academic Press, Inc., San Diego, California (1990).

Gray, R. M. and Davisson, L. D., *Random Processes – A Mathematical Approach for Engineers*. Prentice-Hall, Inc., Englewood Cliffs, New Jersey, USA (1986).

Gubner, J. A., Probability and random processes for electrical and computer engineers (2006).

Gödel, K., *O Teorema de Gödel e a Hipótese do Contínuo*. Fundação Calouste Gulbenkian, Lisboa, Portugal (1979).

Halmos, P. R., *Naive Set Theory*. D. Van Nostrand Company, Inc., Princeton, USA (1960).

Hashemi, H., Principles of Digital Indoor Radio Propagation. In *IASTED International Symposium on Computers, Electronics, Communication and Control*, pp. 271–273, Calgary, Canada (1991).

Hausdorff, F., *Set Theory*. American Mathematical Society, New York, USA (2005).

Hsu, H. P., *Fourier Analysis (Portuguese)*. Livros Técnicos e Científicos Publishers Ltd., Rio de Janeiro, Brasil (1973a).

Hsu, H. P., *Fourier Analysis (Portuguese)*. Livros Técnicos e Científicos Publishers Ltd., Rio de Janeiro, Brasil (1973b).

James, B. R., *Probabilidade: Um Curso em Nível Intermediário*. Instituto de Matemática Pura e Aplicada – CNPq, Rio de Janeiro, Brasil (1981).

Joyce, J., Bayes' theorem. In Zalta, E. N., editor, *The Stanford Encyclopedia of Philosophy*. Fall 2008 edition (2008).

Kennedy, R. S., *Fading Dispersive Communication Channels*. Wiley-Interscience, New York (1969).

Kolmogorov, A. N. and Formin, S. V., *Introductory Real Analysis*. Dove Publications, Inc., New York, USA (1970).

Lathi, B. P., *Modern Digital and Analog Communication Systems*. Holt, Rinehart and Winston, Inc., Philadelphia, USA (1989).

Lebesgue, H., *Leçons sur l'Integration et la Recherche des Fonctions Primitives*. Gauthier-Villars, Paris, France (1904).

Lecours, M., Chouinard, J.-Y., Delisle, G. Y., and Roy, J., "Statistical Modeling of the Received Signal Envelope in a Mobile Radio Channel". *IEEE Transactions on Vehicular Technology*, 37(4):204–212 (1988).

Lipschutz, S., *Teoria de Conjuntos*. Ao Livro Técnico S.A., Rio de Janeiro, Brasil (1968).

Magalhães, M. N., *Probabilidade e Variáveis Aleatórias*. Editora da Universidade de São Paulo, São Paulo, Brasil (2006).

Mandal, M. and Asif, A., *Continuous and Discrete Time Signals and Systems*. Cambridge University Press, Cambridge, United Kingdom (2007).

Marques, M., *Teoria da Medida*. Editora da Unicamp, Campinas, Brasil (2009).

Mises, R. V., Über die Ganzzahligkeit der Atomgewicht and Verwandte Fragen. *Physikal. Z.*, 19:490–500 (1918).

Neal H. Shepherd, Editor, Received Signal Fading Distribution. *IEEE Transactions on Vehicular Technology*, 37(1):57–60 (1988).

Nedoma, J., The Capacity of a Discrete Channel. In *Transactions of the First Prague Conference on Information Theory, Statistical Decision Functions, Random Processes*, pp. 143–181, Prague, Czechoslovakia. Academia, Publishing House of the Czechoslovak Academy of Science (1957).

Oberhettinger, F., *Tables of Fourier Transforms and Fourier Transforms of Distributions*. Springer-Verlag, Berlin (1990).

Oppenheim, A. V., Willsky, A. S., and Nawab, S. H., *Signals and Systems, Second Edition*. Prentice-Hall of India Private Limited, New Delhi, India (2002).

Papoulis, A., Random Modularion: a Review. *IEEE Transactions on Accoustics, Speech and Signal Processing*, 31(1):96–105 (1983).

Papoulis, A., *Probability, Random Variables, and Stochastic Processes*. McGraw-Hill, Singapore (1991).

Pareto, V., *Cours d'Économie Politique: Nouvelle édition par G.-H. Bousquet et G. Busino*. Librairie Droz, Geneve, Switzerland (1964).

Parzen, E., *Teoría Moderna de Probabilidades y Sus Aplicaciones*. Editorial Limusa S. A., México D. F., México (1979).

Petrov, V. V. and Mordecki, E., *Theoría de la Probabilidad*. DIRAC, Faculdade de Ciências, Universidad de la República, Montevideo, Uruguay (2008).

Phillips, E. R., *An Introduction to Analysis and Integration Theory*. Dove Publications, Inc., New York, USA (1984).

Proakis, J. G., *Digital Communications*. McGraw-Hill Book Company, New York (1990).

Rappaport, T. S., Indoor Radio Communications for Factories of the Future. *IEEE Communications Magazine*, pp. 15–24 (1989).

Rényi, A., *Probability Theory*. Dover Publications, Inc., New York, USA (2007).

Rosenthal, J. S., *A First Look at Rigorous Probability Theory*. World Scientific Publishing Co. Pte. Ltd., Singapore (2000).

Schwartz, M., *Information Transmission, Modulation, and Noise*. McGraw-Hill, New York (1970).

Schwartz, M., Bennett, W., and Stein, S., *Communication Systems and Techniques*. McGraw-Hill, New York (1966).

Spiegel, M. R., *Análise de Fourier*. McGraw-Hilldo Brasil, Ltda., São Paulo (1976).

Struik, D. J., *A Concise History of Mathematics*. Dover Publications, Inc., New York, USA (1987).

Sveshnikov, A. A., *Problems in Probability Theory, Mathematical Statistics and Theory of Random Functions*. Dover Publications, Inc., New York, USA (1968).

Taylor, A., Distributions: What Exactly is the Dirac Delta "Function"?. Internet site, https://www.cantorsparadise.com (2022).

Taylor, A. E., *General Theory of Functions and Integration*. Dove Publications, Inc., New York, USA (1985).

Todd, R., Generalized Function. Mathworld – a wolfram web resource, created by eric w. weisstein, https://mathworld.wolfram.com/GeneralizedFunction.html (2022).

Toussoun, O., "Memoire sur l'Histoire du Nil". *Memoires a l'Institut d'Egypte*, 18:366–404 (1925).

Ventsel, H., *Théorie des Probabilités*. Éditions Mir, Moscow, U.S.S.R (1973).

Weibull, W., A statistical distribution function of wide applicability. *Journal of Applied Mechanics*, 18:293–296 (1951).

Whitrow, G. J., Pierre-simon, marquis de laplace. *Encyclopaedia Britannica*, (18 July 2022) (2022).

Whittle, P., *Probability*. John Wiley & Sons, London, United Kingdom (1970).

Wikipedia contributors, Bijection, injection and surjection — Wikipedia, the free encyclopedia. [Online; accessed 6-May-2022] (2022).

Wilcox, H. J. and Myers, D. L., *An Introduction to Lebesgue Integration and Fouries Series*. Dover Publications, Inc., New York, USA (1994).

Zadeh, L. A., Fuzzy Sets. *Information and Control*, 8(3):338–353 (1965).

Zumpano, A. and de Lima, B. N. B., A Medida do Acaso. *Ciência Hoje*, 34(201):76–77 (2004).

Index

A
Abel, Niels Henrik, 21
Académie des Sciences, 227
Affine
  function, 30
Alexander the Great, 207
Algebra, 6, 10
  Borel, 12
  closure, 11
  De Morgan's, 6
  random variable, 119
Amplifier, 136
Archimedes of Alexandria, 41
Archimedes of Syracuse, 41
Area, 41
Argument
  functional, 79
Aristotle, 42, 207
Ars Conjectandi, 98
AWGN, 156
Axiom
  choice, 3
  Peano, 3
  specification, 3
Axiomatic theory of probability, 114
Axioms, 3
  probability, 105, 106
  Zermelo-Fraenkel, 4

B
Baire, René-Louis, 44, 191

Bayes, Thomas, 109
Beauvais, 44
Belonging, 3
Bernoulli, Jacob, 162
Bernoulli, Jacques, 98, 144, 194
Bessel
  identity, 233
Bienaymé, Irénée-Jules, 193
Bijective
  function, 13, 28, 30
Black hole, 227
Borel
  algebra, 12
  rectangle, 120
Borel, Félix Édouard Justin Émile, 66
Borel, Félix Edouard Juston Émile, 12, 43, 48, 53, 100, 167, 190
Borel-Cantelli's Lemma, 220
Brazil, 38
Bureau des Longitudes, 227

C
Calculus, 42
  fundamental theorem, 42
Cantor
  set, 103
Cantor, Georg Ferdinand Ludwig Philipp, 1, 13, 20, 43
Cardano, Girolamo, 97
Cardinal

number, 11, 13
numbers, 2
Cardinality, 11, 13
Cartesian, product, 39
Cauchy, Augustin Louis, 70
Cauchy, Louis, 42
Cavalieri, Bonaventura, 42
CDF, 122
Central Limit Theorem, 179, 222
    product, 225
CGPM, 46
Chaitin, Gregory, 114
Chernoff, Herman, 200
Chevalier de Méré, 97
Closure, 11
Coefficient
    asymmetry, 130
    correlation, 172
Collegium Mauritianum, 38
Communications theory, 107
Compact
    support, 80
Company
    West India, 39
Complex
    covariance, 189
    expected value, 188
    random variable, 188
    variable
        distribution function, 189
    variance, 189
Complexity
    Kolmogorov, 114
    theory, 114
Complexity theory, 107
Composite function
    impulse, 82
Concept of infinite, 42
Conditional
    moment, 177
Control system, 156
Convergence, 211
    almost sure, 212
    distribution, 214
    in Probability, 212
    mean, 214
    mean of order r, 214
    mean square, 215
    measure, 216
    probability, 212
    relationships, 216
    sure, 213
Correlation, 173
    coefficient, 173
    properties, 174
Counting, 41
Covariance, 172
    complex, 189
Crelle, August Leopold, 20

D

Darboux, Gaston, 51
Darboux, Jean-Gaston, 43
De Ludo Aleae, 97
de Moivre, Abraham, 98, 194
De Morgan's
    algebra, 6
    laws, 6
De Morgan, Augustus, 6
De Ratiociniis in Ludo Aleae, 98
Decreasing
    sequence, 7
Dedekind, Julius Wilhelm Richard, 2
Definite
    integral, 237
Delta
    Dirac, 79
Density function

probability, 109
Derivative
   impulse, 82
Descartes, René, 23, 39
Devil's Staircase, 105
Diagram
   Venn, 3, 110
Differential
   function, 44
Dirac
   delta, 79
Dirac, Paul Adrien Maurice, 69, 73, 94, 138
Dirichlet, Johann Peter Gustav Lejeune, 43, 67
Dirichlet, Lejeune, 52
Discrete
   impulse function, 90
   Unit step function, 90
Disjoint, 4
Distribution
   Bernoulli, 144
   binomial, 145
   chi-square, 139
   conditional, 151
   cumulative, 120
   discrete, 144
   exponential, 122, 135
   geometric, 146
   Laplace, 124, 125
   lognormal, 159
   Nakagami, 159
   Pareto, 161
   Poisson, 148
   probability, 119
   Rayleigh, 156
   Rice, 157
   uniform, 141
   von Mises, 160
   Weibull, 162
Doctrine of Chances, 98
Domain
   function, 27
   relation, 24
Doublet
   function, 85

E

Empty
   relation, 26
Empty set
   probability, 107
Equivalence, 13
   relation, 26
Erlang, Agner Krarup, 162
Estimation error, 156
Euclid of Alexandria, 41
Eudoxus of Cnido, 47
Eudoxus of Cnidus, 41
Euler, Leonhard Paul, 98
Event
   independent, 110
Events
   independent, 113
Expansion
   series, 234
Expectation, 127
   operator, 108
   probability, 108
Expected value, 127, 128
Exponential
   distribution, 135
   function, 87–89

F

Families, 7
Fatou, Pierre Joseph Louis, 205
Fermat, Pierre de, 97

266   Index

Final value
   theorem, 232
Fluxion, 42
Fourier
   transform, 85, 239
Fourier, Jean-Baptiste Joseph, 47
Fractals, 2
Função
   Dirichlet, 52
Function, 28, 31
   affine, 30
   bijective, 13, 28, 30
   characteristic, 149
      two-dimensional, 179
   constant, 63
   definition, 27
   Dirichlet, 65
   domain, 27
   doublet, 85
   exponential, 87–89
   gate, 73
   generalized, 85
   impulse, 73, 81
   indicator, 33, 63
   injective, 30, 33
   integrable, 62
   inverse, 13, 33
   kernel, 79
   Laplace, 75
   many to one, 28
   measurable, 62
   modulus, 62
   one to one, 28
   one-one correspondence, 28
   onto, 28
   probability density
      conditional, 153
   properties, 36
   quantizer, 64
   ramp, 86
   random variable, 132
   range, 27
   simple, 62
   square wave, 71
   step, 62
   surjective, 13, 28, 30
   test, 79
   unit step, 35, 71, 73
Functional, 79
   argument, 79
Fussy set, 35
Fuzzy
   set, 34

G

Galilei, Galilei, 42
Games
   chance, 97
Gate
   function, 73
Gauss
   distribution, 137
Gauss, Johann Carl Friedrich, 164
Gauss, Karl Friedrich, 98
General Conference on Weights and
   Measures, 46
Generalized
   function, 85
Goldschmidt, Carl Wolfgang Benjamin, 165
Gombaud, Antoine, 97

H

Hölder, Ludwig, 203
Heaviside, Oliver, 70
Heinrich Lambert, Johann, 19
Hilbert, David, 22
Holder

inequality, 231
Huygens, Christian, 98

I

Identity
  Bessel, 233
  relation, 26
  trigonometric, 233
Image
  relation, 24
Impulse
  composite function, 82
  derivative, 82
  function, 73, 81
  polynomial, 82
  properties, 81
Impulse function
  discrete, 90
Inclusion, 3
Increasing
  sequence, 7
Indefinite
  integral, 235
Independent
  event, 110
  events, 113
  sets, 113
Indexing, 10
Indicator
  function, 33
Inequality
  Bienaymé, 198
  Hölder, 203
  Holder, 231
  Jensen, 198
  Kolmogorov, 201
  Lyapunov, 204
  Markov, 197
  Minkowsky, 204

  Schwartz, 231
  Schwarz, 203, 217
  Tchebychev, 195
Infimum, 6
Infinite, 13
Information
  theory, 109
Information theory, 107
Injective
  function, 33
injective
  function, 30
Institut National des Sciences et des Arts, 227
Integral, 42
  definite, 237
  existence, 42
  function, 44
  indefinite, 235
  Lebesgue, 65, 105, 127
  Riemann, 42, 48, 51, 79
  Riemann-Stieltjes, 128
  sequence, 62
  Stieltjes, 54, 128
International System of Units, 46
Intersection
  set, 37
Interval
  measure, 101
Inverse
  function, 13, 33
  relation, 26

J

Jacobi, Carl Gustav Jacob, 184
Jensen, Johan Ludwig William Valdemar, 198
Jordan, Marie Ennemond Camille, 43

Journal for Pure and Applied Mathematics, 20

K
Kepler, Johannes, 42
Kernel
    function, 79
Khinchin
    Theorem, 197
Khinchin, Aleksandr Yakovlevich, 106, 197
Kirchhoff, Gustav Robert, 165
Kolmogorov
    complexity, 114
Kolmogorov, Andrei Nikolaevich, 105, 113, 193, 201
Kronecker
    impulse function, 90
Kronecker, Leopold, 2, 21, 90, 203
Kummer, Ernst Eduard, 203
Kurtosis, 130

L
Lagrange, Joseph Louis, 98
Lagrange, Joseph-Louis, 227
Laplace
    function, 75
    transform, 85
Laplace, Pierre-Simon, 109, 226
Laplace, Pierre-Simon de, 75, 124, 193, 222
Laplace, transform, 226
Laurent-Moïse Schwartz, 70
Lavoisier, Antoine-Laurent de, 227
Law
    Large Numbers, 219
    strong, 219
    Weak, 217
Law of Large Numbers
    weak, 217
Lebesgue
    full, 105
    measure, 101
    singular function, 105
Lebesgue, Henri Léon, 44, 54, 57, 66, 101, 191
Lebesgue, Henry Léon, 2
Leibniz, Gottfried Wilhelm, 42, 47, 88
Limit
    inferior, 9
    superior, 8, 9
Load-flow, 179
Lognormal
    distribution, 159
Louis Cauchy, Augustin, 47
Lyapunov, Aleksandr Mikhailovich, 194, 204, 222

M
Maastricht, 39
Machine
    Turing, 114
MacLaurin
    series, 232
Many to many
    relation, 27
Many to one
    function, 28
    relation, 27
Markov
    Theorem, 197
Markov, Andrei Andreyevich, 194
Maxwell, James Clerk, 163
Measurable
    space, 99
Measure, 3, 41, 42, 44
    definition, 57

function, 62
interval, 101
Lebesgue, 57, 67, 101, 119
notion, 43
probability, 3, 100
properties, 101
theory, 44
Median, 130
Method
exhaustion, 41
Method of exhaustion, 42
Minkowsky, Hermann, 204
Mises, Richard Edler von, 105, 114
Moivre, Abraham de, 222
Moment, 172
conditional, 177
Moments, 127, 129, 151

N
Nassau-Siegen, Johan Maurits van, 38
National Institute of Science and Arts, 227
Nebular hypothesis, 227
Netherlands, 38
Network
cognitive wireless sensor, 111
Newton, Isaac, 42, 47
Noise
thermal, 156
Number
infinite, 13
transfinite, 43

O
Office of the Longitudes, 227
One to many
relation, 27
One to one

function, 28
relation, 26
Ordered
pair, 23, 39
Oresme, Nicole, 39

P
Pair
ordered, 23
Pareto, Vilfredo Federico Damaso, 161
Paris Observatory, 227
Parmenides, 42
Partition, 110
Pascal, Blaise, 97
pdf, 122
Peano
axiom, 3
Peano, Giuseppe, 43
Philosophy
rationalism, 39
Poincaré, Jules Henri, 191
Poisson, Siméon, 98
Poisson, Siméon Denis, 163
Poisson, Siméon Denis, 148, 193
Polynomial
impulse, 82
Power system, 156
Power-flow, 179
Price, Richard, 109
Probability, 3
axioms, 105, 106
conditional, 154
density function, 109
empty set, 107
expectation, 108
measure, 100
relative frequency, 105
Probability density function, 122

Probability theory, 107
Probability, measure, 44
Product
    Cartesian, 39
Properties
    impulse, 81

Q
Quantizer
    function, 64

R
Ramp
    function, 86
Random variable, 117
    complex, 188
    function, 132
    sum, 180
Random vector, 189
Range
    function, 27
Rationalism
    philosophy, 39
Rayleigh distribution, 156
Recife, 40
Rectangle, 57
Rectifier, 140
Reflexive
    relation, 26
Relation, 23, 26
    definition, 23
    diagram, 24
    domain, 24
    empty, 26
    equivalence, 13
    equivalent, 26
    graph, 24
    identity, 26
    image, 24

inverse, 26
many to many, 27
many to one, 27
one to many, 27
one to one, 26
reflexive, 26
representation, 24
symmetric, 26
tabular notation, 24
transitive, 26
universal, 26
Relative frequency, 105, 114
Rice
    distribution, 157
Riemann
    integral, 48, 79
Riemann, Bernhard, 47, 53
Riemann, Georg Friedrich Bernhard,
    42, 66, 67, 69, 100
Russia, 113

S
Séries, Taylor, 205
Sample
    space, 108
Schröder-Bernstein
    theorem, 3
Schrödinger, Erwin Rudolf Josef
    Alexander, 70
Schwartz
    inequality, 231
Schwarz
    inequality, 203, 217
Segment, 57
Sequence
    decreasing, 7
    increasing, 7
Series
    expansion, 234

Laurent, 223
MacLaurin, 232
trigonometric, 42
Set, 1
  algebra, 6
  Cantor, 103
  continuous, 43
  conventional, 35
  countable, 43
  disjoint, 4
  empty, 3
  families, 7
  fuzzy, 34
  infinite, 2, 43
  integer numbers, 15
  internal, 43
  intersection, 37
  measurable, 57
  measure, 44
  natural numbers, 15
  operations, 5
  points, 43
  rational numbers, 15
  real numbers, 15
  rectangle, 57
  union, 36
  universal, 3
  universal set, 1
Set theory, 1, 2
Sets
  algebra, 10
  independent, 113
  indexing, 10
Shannon, Claude Elwood, 114
SI, 46
Singer, Georg Ferdinand Ludwig Philipp, 103
Singular function
  Lebesgue, 105

Sobolev, Sergei Lvovich, 70
Solar System, 227
Solomono, Ray J., 114
Space
  measurable, 99
  sample, 108
Spain, 39
Square
  wave, 77
Square wave
  function, 71
Statistical Expectancy, 172
Stevin, Simon, 39
Stieltjes, Thomas Joannes, 54
Stochastic processes, 107
Support
  compact, 80
Supremum, 6, 37
Surjective
  function, 13, 28, 30
Symmetric
  relation, 26
Syracuse, 41

T

Tambov, 113
Tchebychev
  Theorem, 196
Tchebychev, Pafnuty Lvovich, 193
Test
  function, 79
Théorie Analytique des Probabilités, 227
Theorem
  Bayes, 109
  final value, 232
  fundamental, 42
  Khinchin, 197
  Markov, 197

Schröder-Bernstein, 3
Tchebychev, 196
Theory
  complexity, 114
  information, 109
  measure, 44
  set, 44
Theory of probability, 105
  axiomatic, 114
Timon of Phlius, 207
Transfinite arithmetic, 2
Transform
  Fourier, 85, 239
  Laplace, 85
Transform, Laplace, 226
Transformation
  general formula, 138
  uniform, 143
Transitive
  relation, 26
Trigonometric
  identity, 233
Turing's
  machine, 114
Turing, Alan, 114

U

Union
  set, 36
Unit step
  function, 35, 71, 73
Unit step function
  discrete, 90
Universal
  relation, 26

Universal set, 3
University of Halle, 2

V

Variance, 129
  complex, 189
Venn
  John, 4
  diagram, 3, 110
Volterra, Vito, 18, 51
von Mises
  distribution, 160

W

Wave
  square, 77
Weber, Wilhelm Eduard, 164
Weibull, Waloddi, 162
Weierstrass, Karl Theodor Wilhelm, 203
West India
  Company, 39

Z

Zadeh, Lotfi Aliasker, 34
Zeno, 1
Zeno of Elea, 42
Zermelo-Fraenkel
  axioms, 4
Zorn
  lemma, 3
Zorn's lemma, 3

About the Authors

Marcelo S. Alencar was born in Serrita, Brazil in 1957. He received his Ph.D. from the University of Waterloo in 1994. He has more than 40 years of engineering experience, and he is an IEEE Senior Member. He has been involved with consulting and project development for many companies, agencies, and universities. He is Visiting Professor at the Department of Communications Engineering (DCO), Federal University of Rio Grande do Norte, and a former Full Professor at the Federal University of Campina Grande, Brazil. He spent sabbatical leaves working for Embratel and University of Toronto.

He is founder and President of the Institute for Advanced Studies in Communications (Iecom). He has been awarded several fellowships and grants. He holds the award for achievement from the College of Engineering of the Federal University of Pernambuco. He was honored by the Federal University of Paraiba. He received the prestigious Attilio Giarola Medal from the Brazilian Microwave and Optoelectronic Society (SBMO). He published over 550 engineering and scientific papers. He authored 30 books and wrote chapters for twelve books.

About the Authors

Marcelo S. Alencar is editor of the River Publishers Series in Signal, Image and Speech Processing and has contributed in different capacities to the following scientific journals: Editor of the Journal of the Brazilian Telecommunication Society (SBrT); Founder of the Journal of Communications and Information Systems (SBrT); Member of the International Editorial Board of the Journal of Communications Software and Systems (JCOMSS), published by the Croatian Communication and Information Society (CCIS); Member of the Editorial Board of the Journal of Networks (JNW), published by Academy Publisher; Co-Editor-in-Chief of the Journal of Communication and Information Systems, jointly sponsored by the IEEE Communications Society (ComSoc) and SBrT. He was member of the IEEE, SBMO and SBrT Boards. He is a Registered Professional Engineer and recipient of two grants from the IEEE Foundation.

He published the following books: *Nanotechnology-Based Smart Remote Sensing Networks for Disaster Prevention*, by Elsevier, *Economic Theory, Cryptography and Network Security, Music Science, Linear Electronics, Modulation Theory, Scientific Style in English*, and *Cellular Network Planning*, by River Publishers, *Spectrum Sensing Techniques and Applications, Information Theory, and Probability Theory*, by Momentum Press, *Information, Coding and Network Security* (in Portuguese), by Elsevier, *Digital Television Systems*, by Cambridge, *Communication Systems, 3rd Edition*, by Springer, *Principles of Communications* (in Portuguese), by Editora Universitária da UFPB, *Set Theory, Measure and Probability, Computer Networks Engineering, Electromagnetic Waves and Antenna Theory, Probability and Stochastic Processes, Digital Cellular Telephony, 3rd Edition, Digital Telephony, 5th Edition, Digital Television and Communication Systems* (in Portuguese), by Editora Érica Ltda, *History of Communications in Brazil, History, Technology and Legislation of Communications, Connected Sex, Scientific Diffusion, Soul Hicups* (in Portuguese), by Epgraf Gráfica e Editora. He also wrote several chapters for 14 books.

Marcelo S. Alencar has contributed in different capacities to the following scientific journals: Editor of the *Journal of the Brazilian Telecommunication Society*; Member of the International Editorial Board of the *Journal of Communications Software and Systems* (JCOMSS), published by the Croatian Communication and Information Society (CCIS); Member of the Editorial Board of the *Journal of Networks* (JNW), published by Academy Publisher; founder and editor-in-chief of the *Journal of Communication and Information Systems* (JCIS), special joint edition of the IEEE

Communications Society (ComSoc) and SBrT. He is a member of the SBrT-Brasport Editorial Board. He has been involved as a volunteer with several IEEE and SBrT activities, including being a member of the Advisory or Technical Program Committee in several events. He served as a member of the IEEE Communications Society Sister Society Board and as liaison to Latin America Societies. He also served on the Board of Directors of IEEE's Sister Society SBrT. He is a Registered Professional Engineer. He was a columnist of the traditional Brazilian newspaper Jornal do Commercio for two decades, and he was vice-president external relations of SBrT. He is a member of the IEEE, IEICE, in Japan, SBrT, SBMO, SBPC, ABJC and SBEB, in Brazil. He studied acoustic guitar at the Federal University of Paraiba, and keyboard and bass at the music school Musidom. He is the perpetual president, composer and percussionist of the carnival club *Bola de Ferro*, in Recife, Brazil.

Raphael T. Alencar was born in Recife, Brazil, in 1988. He obtained his Bachelor Degree in Electrical Engineering Degree (Majors in Telecommunications and Power Systems), in 2012, and his Master¿s Degree in Electrical Engineering (Telecommunications) both from the Federal University of Campina Grande, Brazil, in 2014. He has a Ph.D. from the University of Grenoble, France. He also got involved in an International exchange program at the École Nationale Supérieuse des Systèmes Avancés et Réseaux, ES-ISAR, Valence, France, from 2010 to 2011. He now works for the company Raydiall Automotive, Voiron, France.

He worked as a consulting engineer in a project for Alpargatas (Havainas© manufacturer and distributer in the country), and his main activities included development, manufacture, installation and testing of a security system to protect mill operators. The main tasks involved electronic and mechanic designs, image processing, sensors and radiofrequency systems testing. He also worked at LCIS ¿ Laboratoire de Conception et d¿Intégration des Systèmes, in the development and manufacture of a chipless RFID reader.

He was a researcher in a Project funded by the Brazilian National Research Network (RNP): "$[CIA]^2$ – Building Intelligent Cities: from Enviromental Instrumentation to Application Development", developed at the Institute of Advanced Communications Studies (Iecom). His activities included mostly research, design and simulation for optical communications and hybrid optical-wireless networks. He is now with Raydiall Automotive, Voiron, France, a joint venture between ARaymond, a world leader in the automotive fastening industry, and Radiall, a world leader in interconnection solutions.